Blue light wavelength: 1.9×10^{-5} inch

Interstellar dust grain: diameter 4×10^{-6} inch

Cell: diameter 5×10^{-4} inch

Black hole: diameter 40 miles

Large moon crater: diameter 120 miles

Largest asteroid: diameter 620 miles

Mars: diameter 4,223 miles

White dwarf: diameter 5,000 miles

Venus: diameter 7,521 miles

THE NEAR PLANETS

A far-flung ocean of gas and dust spanning billions of miles collapses into the Solar nebula, birthplace of the planets.

Over time, rocky fragments form from the nebula's dust, colliding and combining in the pallid light of the proto-Sun.

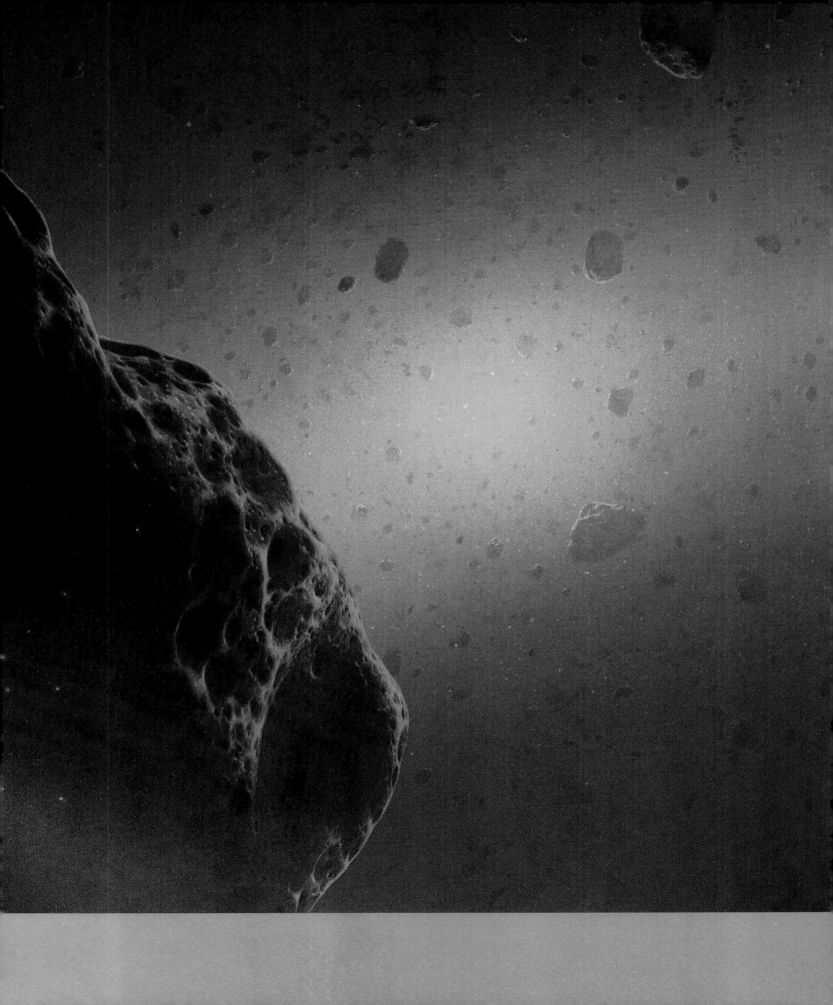

Gravity collects smaller rocks into protoplanets that sweep up nebular debris as they orbit the nascent Sun.

Compressed by in-falling gas, the Sun ignites in a blast of solar wind, stripping nearby planets of their atmospheres.

In a brutal hail of meteoroids, an inner planet rebuilds its atmosphere from the gases of erupting volcanoes.

This volume is one of a series that
examines the universe in all its aspects,
from its beginnings in the Big Bang to the
promise of space exploration.

VOYAGE THROUGH THE UNIVERSE

THE NEAR PLANETS

BY THE EDITORS OF TIME-LIFE BOOKS
ALEXANDRIA, VIRGINIA

CONTENTS

Earth's nearest neighbors in the Solar System—airless Mercury, sulfurous Venus, and the ice-capped desert world of Mars—are all smaller than the third planet.

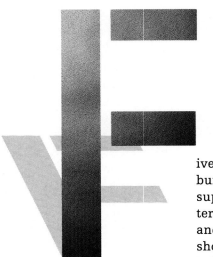

ive billion years ago, a star died in a silent, blinding burst of fury. Shock waves ballooned outward from the supernova. Racing through the galaxy, they encountered a neighbor, a vast rotating cloud of gas and dust, and drastically altered its fate. Compressed by the shock wave, the cloud's gas molecules—hydrogen and helium—began to cling together. As the cloud spun slowly about, gravity pulled the agglomerations into an increasingly dense center. The cloud's outer regions flattened out, and the impact of in-falling gas heated the glowing core. After a few thousand years, the cloud became an enormous, spinning disk. Known as the Solar nebula, it was the ancestor of the planets in today's Solar System.

Slowly the nebula began to condense, bit by bit, into a solar family. Continuing to be fed by falling matter, the massive central bulge grew into a hot, spherical proto-Sun. Clumps of dust circling closest to it became pebbles of iron, silicon, magnesium, aluminum, nickel, and other metallic elements. Meanwhile, in the cold outer reaches of the nebula, lighter, more volatile compounds such as water and methane began to precipitate out of their gaseous state, much as earthly frost forms on a cold night. Pulled by gravity toward the central plane of the disk, the glittering fragments collected in whirlpools and eddies. Collisions were frequent, and gradually the debris grew into asteroid-size bodies known as planetesimals.

Now planet building began in earnest. Thousands of planetesimals gathered in loose clusters, marking out distinct orbits around the proto-Sun as they coalesced into massive planetary bodies. Rock strewn, turbulent, and veiled in gas and dust, the primitive solar system was ready for a rite of passage: the birth of its star.

When the crushing weight of accumulated hydrogen reached a critical mass, a thermonuclear reaction called fusion exploded inside the young Sun. During this so-called T Tauri phase (the name comes from the star in the constellation Taurus where the phenomenon was first identified), the Sun blasted away its outermost layers, propelling them through the fledgling solar system. The T Tauri wind stripped the inner planets of their primordial atmospheres and cleared out most of the remaining gas and dust from the Solar nebula.

Farther from the blast point, the outer worlds managed to hang onto dense, gaseous envelopes of hydrogen and helium, but the inner planets were reduced to barren spheres of rock and metal. Heat from the decay of heavy, radioactive elements melted the interiors of the rocky worlds, allowing iron-rich minerals to sink to the planetary cores and forcing lighter silicates to the surface. Gases escaped from the molten interiors through vents and volcanoes and formed new atmospheres composed mainly of carbon dioxide and nitrogen. For half a billion years more, the young planets endured an intense bombardment by chunks of unconsolidated debris left over from the planet-forming era. This deadly rain ended some four billion years ago, but not before gouging the planetary surfaces, leaving behind scars that still bear witness to the violent birth of the Solar System.

Their formation completed, the inner four members of the Sun's family—Mercury, Venus, Earth, and Mars—started down separate roads. Collectively known as the terrestrial planets, the quartet now offers starkly contrasting examples of the fates that can befall closely related worlds.

ALIEN LANDSCAPES

Billions of years after the inner planets took shape, robotic invaders appeared in their skies. Beginning in the 1960s, compact probes bristling with detection equipment were launched from the third planet. Some reconnoitered and landed on Venus and Mars, and within a decade, another had repeatedly brushed past Mercury. The space age had arrived: Earth's dominant life form was exploring the planetary neighborhood.

For centuries, astronomers had attempted to understand the Solar System through the telescope. Unlike the distant stars, never more than points of light through the lens, planets were relatively close by and easily identified as worlds. Scientists were particularly interested in knowing whether the inner planets resembled the Earth. Had they developed Earth-like atmospheres or oceans—or even life?

Yet even the largest lens and the most sophisticated analyses of reflected light failed to answer the needs of planetary science. Mercury was too distant and too well hidden in the Sun's glare for its surface to be seen. Venus was shrouded in clouds. The details visible on the face of Mars produced more controversy than conclusions. To truly understand the neighboring planets, astronomers had to send their instruments out there, across the millions of miles, on the backs of spacecraft. Many failed to reach their targets, but those that did manage to complete their journeys—the Viking missions to Mars, Venera to Venus, Mariner to Mercury, and others—began at last to solve some of the longstanding puzzles.

None of the modern miracles of interplanetary flight would be possible were it not for the scrupulous observers and logical thinkers of an earlier era. Before astronomers knew what the planets were, they had figured out where they were and how they moved. Working without telescopes, sixteenth-century astronomers deciphered the mysteries of celestial mechanics by sheer brainpower. In 1543, Polish Catholic cleric Mikolaj Kopernik, better known as Nicolaus Copernicus, published a tract that broke with centuries of accepted wisdom by affirming that the motions of the planets could best be explained if all bodies orbited the Sun rather than the Earth. Following up on his controversial theory at the turn of the century was Johannes Kepler, a dreamy and impoverished German schoolteacher. Kepler realized that Copernicus's brilliant insight, which postulated perfectly circular orbits, still did not account for certain irregularities that Kepler saw in planetary movement. After years of theoretical work based on new planetary observations by his colleague Tycho Brahe, he decided that the planets must follow elliptical orbits around the Sun—curves mathematically defined by their distance from two fixed focal points *(opposite)*.

This discovery—which came to be known as Kepler's first law—was one of the great breakthroughs in the study of astronomy. Kepler could then formulate other laws of planetary motion, which neatly described the movements, speed, and orbital characteristics of all bodies in motion around the Sun *(page 35)*. Within eighty years, Isaac Newton would round out the revolution in astronomy by demonstrating that these laws were the consequences of universal gravitation.

Kepler showed how the planets move, and Newton would ultimately explain why they move, but the bodies themselves remained enigmatic. Help was not long in arriving. Even as Kepler was publishing his laws of motion, the Italian mathematician and physicist Galileo Galilei turned his first telescope toward the skies.

Galileo was a lecturer in mathematics at the University of Padua when he heard of the new device, already in use in England and elsewhere. Immediately he constructed his own greatly improved instrument and, in so doing, transformed the science of astronomy. Everywhere he looked, Galileo saw new wonders in the heavens—thousands of previously unseen stars, craters and mountains on the Moon, satellites around Jupiter, and what turned out to be rings around Saturn. By the end of the century, astronomers had begun to compile a body of knowledge about each of the nearest planets. Mars proved most intriguing.

THE RIDDLE OF MARS

Revealed through the telescope, Mars turned out to be a rich source of both information and speculation. In 1659, Dutch scientist Christiaan Huygens recorded the existence of a dark, triangular area on the Martian surface. Later named Syrtis Major, this distinctive feature allowed Huygens to calculate how long the planet took to spin once on its axis. "The rotation of Mars, like that of Earth," he observed, "seems to be in a period of twenty-four hours." About seven years later, the Italian astronomer Giovanni Domenico Cassini made more precise observations and came up with a length for the Martian day of twenty-four hours and forty minutes, a value that overestimated the actual rotational period by just two and a half minutes. Cassini and his assistant, Jean Richer, also made the first calculation of the distance to Mars. In 1671, Cassini, in Paris, and Richer, in South America, each recorded Mars's position

Exploring the Ellipse

Early in the seventeenth century, Johannes Kepler discovered that the orbital path of every planet in the Solar System is an oval-shaped geometric figure called an ellipse. Two points, known as foci, define the figure according to a simple rule: The sum of the distances from each of the two foci to any point on the ellipse is always the same. In the diagram of a planetary orbit at right, the Sun is at one focus and a point in space is the second focus. The combined length of the lines drawn in blue from the Sun and from the other focus to a planet located at orbital point A is identical to the sum of the red lines connecting the two foci to the orbital point labeled B.

A line that passes through both foci and extends to either side of the ellipse is called the major axis, represented here by a solid white line. The endpoints of this axis are called vertices. An orbiting planet at the vertex farthest from the Sun is said to be at aphelion, from the Greek for "off from the Sun" (C); the point of closest approach is called perihelion, "close about the

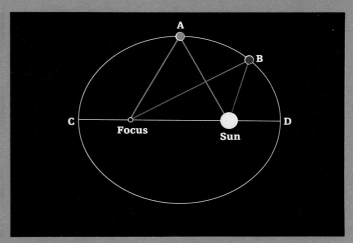

Sun" (D). In the special case of an ellipse that is a circle, the two foci are superimposed at center, halfway between aphelion and perihelion. Otherwise, the foci occupy distinct positions; the greater the separation between them, the more elongated, or "eccentric," a planet's orbit.

relative to the background stars. Combining the differences between the measurements with the known distance between the two observers, the scientists used simple geometry to calculate a distance to Mars of some 40 million miles. With this figure—roughly equal to the diameters of 5,000 Earths—the vast scale of the Solar System began to be apparent.

By the 1780s, astronomers such as German-born William Herschel had focused their attention on what was to become the main issue in Martian astronomy: life. The planet's surface was mottled with bright and dark patches. Herschel and others took these to be contrasting areas of land and sea. Herschel also detected the existence of "a considerable but moderate atmosphere" on Mars. Noting that the appearance of the Martian surface seemed to change over time, he speculated that these changes could be ascribed to the effects of "clouds and vapors floating in the atmosphere of the planet." All things considered, Herschel concluded, "the inhabitants [of Mars] probably enjoy a situation similar to our own."

As advancing technology permitted the construction of larger and better telescopes throughout the nineteenth century, the quality of Martian maps improved dramatically. Nevertheless, detailed observation remained difficult. One reason is that Mars's small size makes it hard to see clearly even under the best of conditions. A more telling reason is that the planet is almost invisible for prolonged periods when it is on the other side of the Sun from the Earth. Every 780 days, however, the orbits of the two planets bring them to a minimum distance from each other, in an alignment known as opposition. Every fifteen and seventeen years, that is, in pairs, two years apart, an exceptionally close opposition occurs when Mars and Earth may be separated

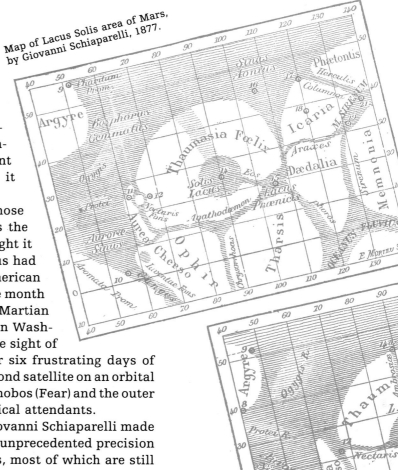

Map of Lacus Solis area of Mars, by Giovanni Schiaparelli, 1877.

The same region, as Schiaparelli observed it in 1879.

by no more than 35 million miles. One of these favorable oppositions took place in 1877, and astronomers grabbed the opportunity to make two important discoveries—one of them true and the other, as it turned out, spectacularly false.

The first was the detection of Martian moons, whose possible existence had been suggested as early as the seventeenth century by Johannes Kepler (who thought it only logical that Mars should have two, since Venus had none and Earth had one). A forty-eight-year-old American astronomer named Asaph Hall decided to devote the month of August 1877 to making a systematic search for Martian satellites. Working at the U.S. Naval Observatory in Washington, D.C., he was rewarded on August 11 with the sight of one small object moving around Mars. Then, after six frustrating days of foggy weather, the skies cleared and he spotted a second satellite on an orbital path inside of the first. Hall named the inner moon Phobos (Fear) and the outer moon Deimos (Panic) after the war god's mythological attendants.

During that same summer, Italian astronomer Giovanni Schiaparelli made a detailed study of Mars, mapping its surface with unprecedented precision and assigning to it a brand new set of place names, most of which are still in use today. But Schiaparelli is mainly remembered for his description of a number of long, thin, straight lines crisscrossing the Martian surface. Other astronomers had seen some of these features, but Schiaparelli mapped them and named them *canali,* an Italian word meaning "channels." However, in an age of canal building on Earth (the Suez Canal had opened in 1869), it was probably inevitable that Schiaparelli's *canali* would be likened to the man-made versions.

The idea of Martian canals gripped the popular imagination, and particularly that of a rich American named Percival Lowell. In 1894, at the age of thirty-nine, he established the Lowell Observatory at Flagstaff, Arizona, where, although he had no formal scientific training, he devoted himself to the study of Mars and especially its network of canals.

By this time, many astronomers had come to doubt that there were, as Herschel had proposed, oceans and seas on Mars. Large bodies of standing water ought to reflect sunlight in brilliant flashes, but none had been seen. The dark areas on the Martian surface—which looked green to some eyes— were now widely believed to be regions of vegetation, literal oases in the midst of a rust-red desert that covered much of the planet. Some astronomers had observed a "wave of darkening" that seemed to move from the poles toward the equator during each hemisphere's spring, as if melting ice were providing water for seasonal growth. To Lowell, it was clear that the water needed some kind of delivery system, and the canals, many of which seemed to connect isolated dark regions, were the logical candidates.

In three widely read books, Lowell laid out his case for a Martian civilization. He believed that Mars was a dry, dying world where intelligent beings labored heroically to survive in an increasingly hostile environment. "A mind of no mean order would seem to have presided over the system we see," Lowell wrote. "Certainly what we see hints at the existence of beings who are in advance of, not behind, us in the journey of life."

Thrilling as it was to the public, Lowell's vision was a source of annoyance to professional astronomers. Although he claimed to have seen some 500 canals, most members of the astronomy community never glimpsed any. And even if the canals did exist, it seemed more likely that they were some sort of natural phenomenon. Modern planetary scientists now find the whole canal controversy a historical embarrassment, generally believing that the canals were simply an optical illusion produced by the mind's eye, artifacts of a brain eager to impose order on disconnected surface details.

DESERT WORLD

His Martian obsession notwithstanding, Lowell went on to do other solid planetary work at Flagstaff, as did members of his staff. They and other astronomers of the time found that some of their questions about Mars could be answered not with the eye but with the science of spectroscopy, which became a powerful research tool in the nineteenth and twentieth centuries.

Spectroscopy is the analysis of information carried in light. In the nineteenth century, long after the English physicist Isaac Newton discovered that a prism could split sunlight into a spectrum of colors, other scientists found that the colored array was subtly striped with thousands of narrow, dark lines. Researchers deduced that as light passed through the solar and earthly atmospheres, chemical elements in those atmospheres absorbed certain wavelengths, producing gaps in the spectrum. By comparing the absorption lines with those produced by specific gases in the laboratory, scientists could pick out elements such as sodium and oxygen in the Sun. Taking the procedure one step further, astronomers were able to study planets as well: They compared the spectrum of light reflected from a planet with the solar spectrum and found absorption lines characteristic of that world. Then they matched the lines against their laboratory spectra to identify each element.

The task was a tricky one. Scientists needed to separate lines created by Earth's own atmosphere from those of the Sun or planets. Their indispensable aid in this task was the physical effect known as the Doppler shift. Named after Austrian physicist Christian Doppler, this effect—an apparent frequency change—shows up when wavelengths of light from the object being observed are compressed or stretched as the object moves toward or away from the observer. The shift makes it just possible to isolate the spectrum of another planet or the Sun. Because the Sun is rotating and other planets are moving relative to Earth, their absorption lines appear

Schiaparelli's 1881 map of Lacus Solis and its surroundings.

23

in a slightly different part of the spectrum from those of the home planet.

By the 1930s, scientists had determined with fair certainty that the Martian atmosphere contained little or no oxygen or water vapor. In 1947, Dutch-American astronomer Gerard Kuiper detected the spectral lines of carbon dioxide and identified it as the major constituent of the atmosphere of Mars. Far from indicating a benign, watery world, such readings showed that Mars would be coldly hostile to Earth-like life. All of this was bad news for believers in Martian civilizations.

Twentieth-century astronomers left the canals behind and relegated tales of Martian agriculture to the realm of science fiction. By the 1960s, observations and photographs of the Martian surface seemed to show vast, planetwide dust storms. The movement of these storms might explain the appearance and disappearance of dark regions on the planet, thus ruling out seasonal vegetation. Measurements of the planet's infrared light—the invisible radiation given off by warm objects—showed that even in the daytime, temperatures were usually well below the freezing point of water. Spectroscopic observations of crisp-looking CO_2 absorption lines from Mars showed that the gas was under little pressure. (Gas under high pressure creates blurred lines in the spectrum.) From this, astronomers estimated the planet's atmospheric pressure at about 85 millibars (a millibar measures pressure per square centimeter), versus about 1,000 millibars on Earth. In short, Mars appeared to be dry and frigid, with a thin carbon dioxide atmosphere insufficient to support any but perhaps the most primitive forms of life.

The search for water and life was not limited to the Red Planet. In fact, the Lowellian canals of Mars had their counterpart in the steaming tropical swamps of Venus. Because Venus is the closest planet to Earth (some 24 million miles away, at its nearest approach), as well as the closest in size, some people reasoned that it was most like Earth. But unlike Mars, Venus hid under an impenetrable cover of clouds.

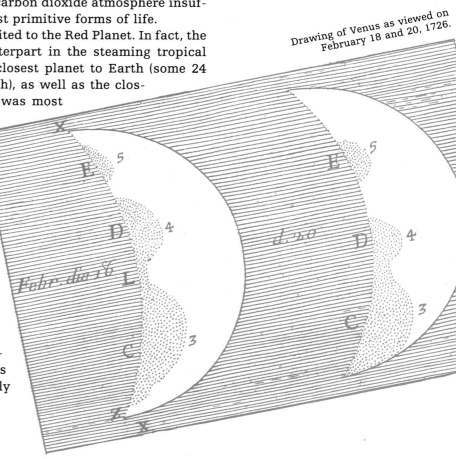

Drawing of Venus as viewed on February 18 and 20, 1726.

THE LURE OF VENUS

One of the few early observations of Venus of any significance came soon after the telescope was invented. In 1610, Galileo discovered that the brilliant planet of the morning and evening skies had phases: Its illuminated shape waxed and waned like that of the Moon, going from crescent to gibbous to full and back again. The news was stunning, since conventional wisdom held that both Venus and the Sun orbited the Earth. In the orthodox view of orbital relationships, Venus's unilluminated hemisphere would continually

Venus as sketched by an English
astronomer in 1882.

face the Earth, and Venus would always resemble a crescent Moon. Phases could occur only if both Venus and the Earth were orbiting the Sun *(pages 36-37)*.

Galileo spread word of his discovery cautiously at first, through coded letters to colleagues, but he soon gained confidence and announced it publicly, helping to establish the Copernican system. Yet even with a more accurate picture of celestial movement and with rapidly improving telescopes, little new information was gathered about the stubbornly mysterious planet in succeeding centuries. Twice in the 1760s, Venus passed before the Sun's disk in what is called a transit. During the first, in 1761, Russian astronomer Mikhail Lomonosov noted that the planet's outline was blurred and hazy, leading him to suspect the existence of a Venusian atmosphere. Its presence was confirmed by German astronomer Johann Hieronymus Schröter, who worked at his own observatory near Bremen from 1778 until 1814, when Napoleon's invading troops plundered his brass-tubed telescope because they thought it was gold. Schröter noted that the "horns," or cusps, of the crescent Venus sometimes seemed to extend beyond a semicircle, suggesting that sunlight was being diffused by a thick atmosphere.

Interesting as it was, the planet's opaque and virtually featureless cloud cover remained a frustrating obstacle, preventing astronomers from determining even something so basic as Venus's period of rotation. Most early estimates put the Venusian day at about twenty-four hours, a figure that was off target by some eight months. Into this informational void rushed pure speculation. In 1918, for example, Swedish scientist Svante Arrhenius offered a description of the probable conditions on the second planet. Arrhenius had been a harsh critic of Lowell's Martian assertions, yet—like Lowell—he extrapolated from a bare minimum of data: "The average temperature there is calculated to be about 47° C," he wrote. "The humidity is probably about six times the average of that of Earth, or three times that in the Congo, where the average temperature is 26° C." He concluded furthermore that "a very great part of the surface of Venus is no doubt covered with swamps, corresponding to those on the Earth in which the coal deposits were formed." The temperature and humidity, he averred, promoted "a luxuriant vegetation." The image of Venus as a primitive swamp world made for a logical continuum: young, steamy Venus; temperate, middle-aged Earth; and cold, dry, ancient Mars.

Repeated attempts to verify such speculations were in vain. The oxygen and water vapor in Earth's own atmosphere interfered with the spectroscopic study of light reflected from Venus, making the substances extremely difficult to detect in the other planet's atmosphere, despite the Doppler shift. Because

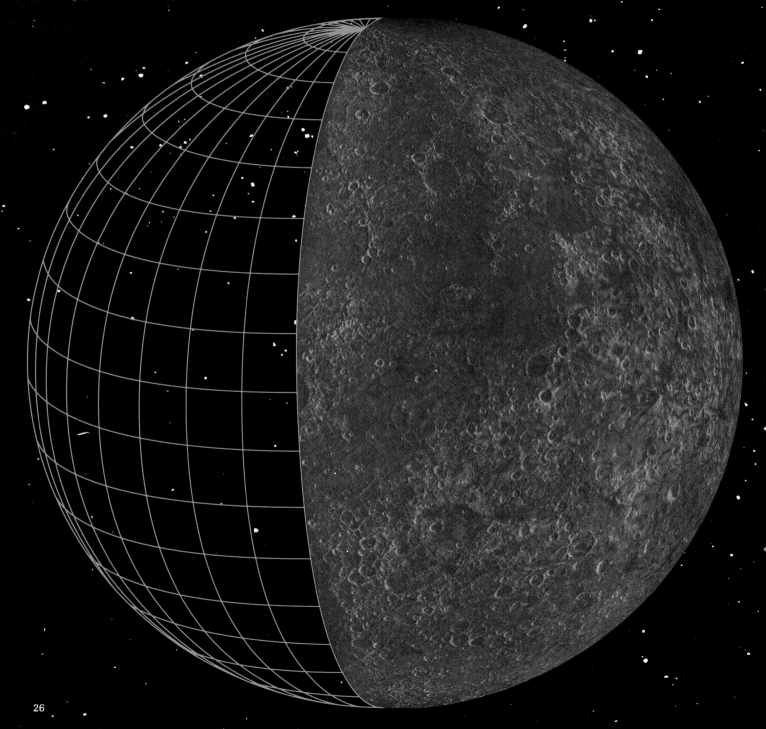

Mapping Other Worlds

Although no human has yet traveled across the inhospitable landscapes of Mercury, Venus, or Mars, surprisingly detailed maps of these alien worlds exist.

Prepared for NASA by the U.S. Geological Survey's Branch of Astrogeology in Flagstaff, Arizona, they not only illuminate the planets' surface features but yield clues to their geologic histories as well.

Hundreds of maps are available for Mars alone, derived from the pictorial bounty returned in the 1970s by *Mariner 9* and *Viking*s *1* and *2*. A score depict Mercury, visited by *Mariner 10* in 1974-1975. Venus, shrouded in opaque clouds, can only be viewed by radar. While circling the planet from 1978 through 1980, NASA's Pioneer Venus Orbiter bounced radar pulses off the surface to determine the height of the world's larger features. Scientists on Earth used computers to analyze the data and transform it into vividly colored maps. In 1983 and 1984, two orbiters from the then-Soviet Union, *Venera*s *15* and *16*, traced the contours of much of Venus by a technique called synthetic-aperture radar (SAR). Unlike the Pioneer method, SAR uses a broad, fanlike signal that yields enough data to produce an impressively detailed portrait of the planet. More recently, Venus has been examined by the *Magellan* spacecraft, whose advanced SAR has allowed it to return photograph-like images of unprecedented clarity and resolution *(pages 80-83).*

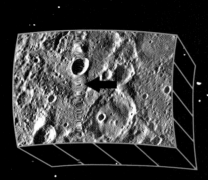

The tiny crater Hun Kal *(above, indicated by arrow, and outlined at right)* lies directly on Mercury's twenty-degree line of longitude and is named for the ancient Mayan word for "twenty." By a 1970 agreement among astronomers, a line that passed through the point calculated to be closest to the Sun at perihelion in the year 1950 was declared the zero-degree longitude. When *Mariner 10* photographed the planet twenty-four years later, however, that area was in shadow. Needing to anchor their longitude system to a landmark, scientists selected a nearby visible crater—little Hun Ka —as the point through which the twentieth meridian passes.

Mariner 10 photographed the area at right as it approached Mercury and the area at left after it passed by. Cartographers used the images, taken when the spacecraft was between 100,000 and 200,000 miles away, to create the two photomosaics, which together cover about half the planet. After fixing longitude and latitude with respect to Mercury's spin axis, the mapmakers determined the coordinates of the planet's surface features.

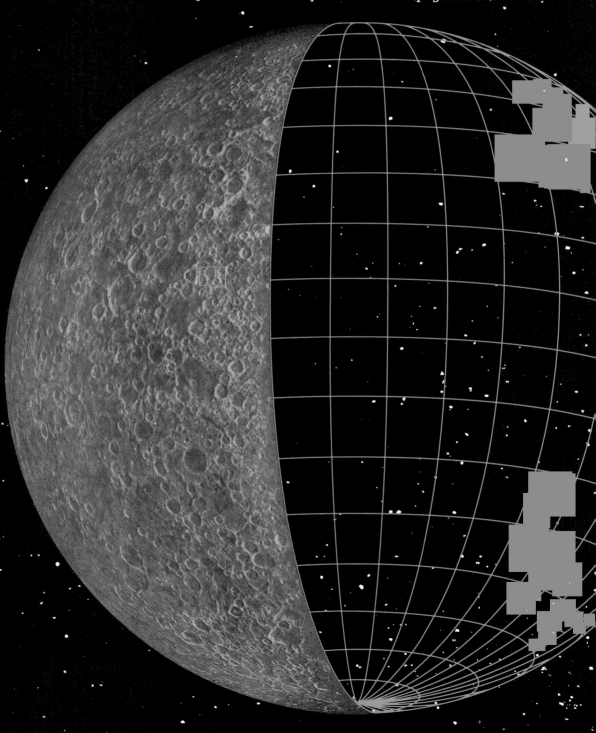

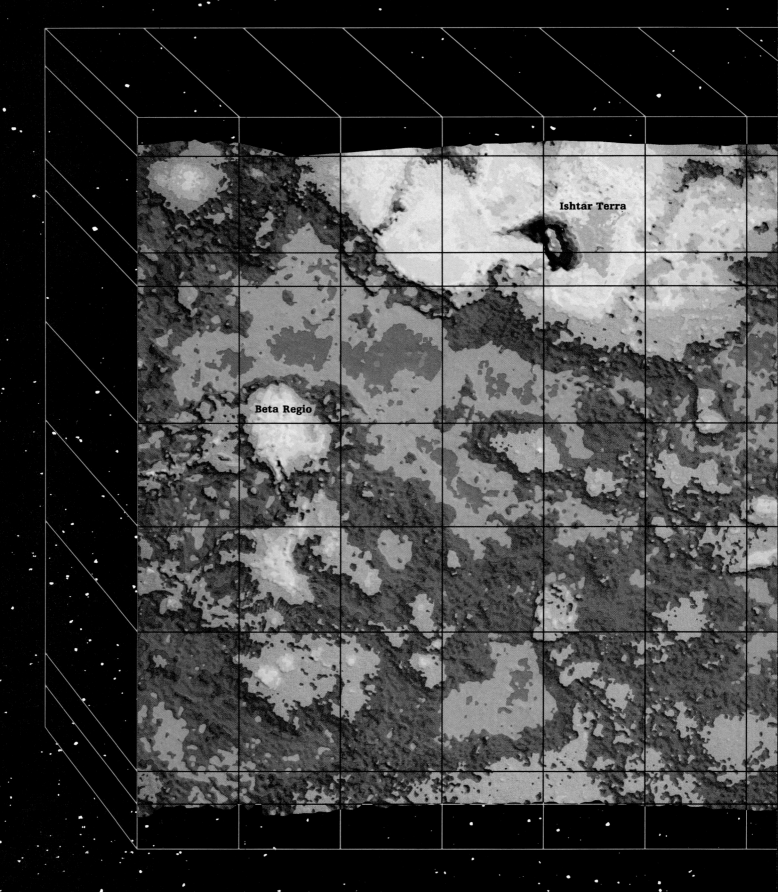

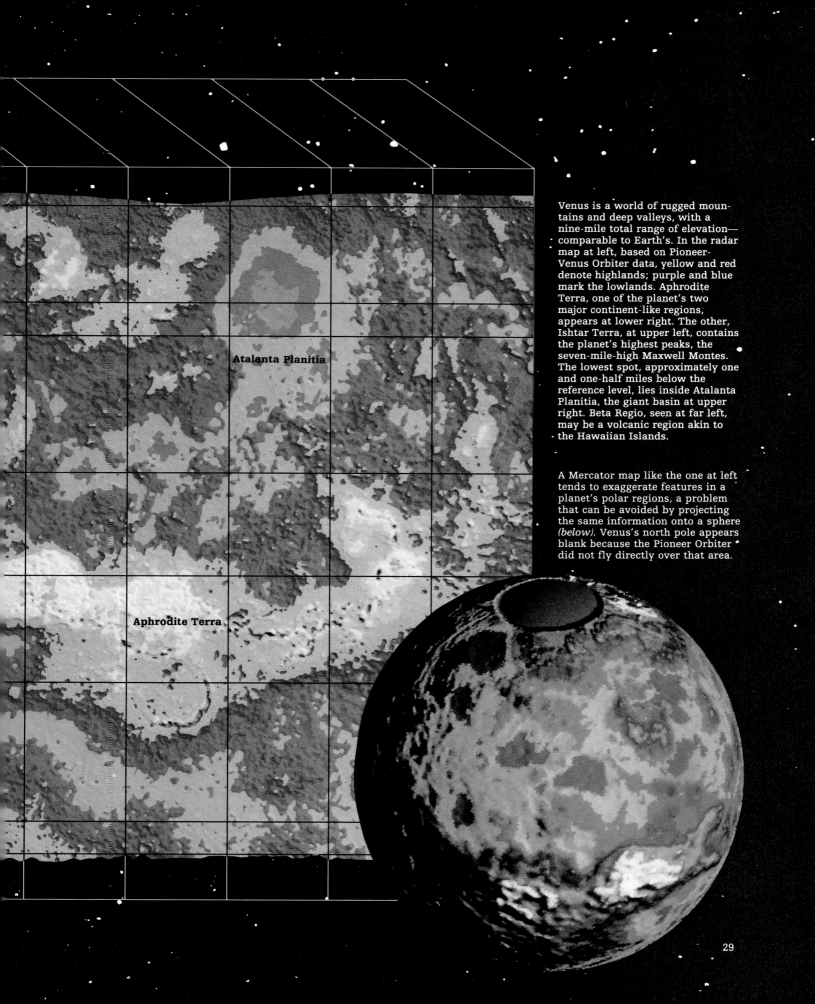

Atalanta Planitia

Aphrodite Terra

Venus is a world of rugged mountains and deep valleys, with a nine-mile total range of elevation—comparable to Earth's. In the radar map at left, based on Pioneer-Venus Orbiter data, yellow and red denote highlands; purple and blue mark the lowlands. Aphrodite Terra, one of the planet's two major continent-like regions, appears at lower right. The other, Ishtar Terra, at upper left, contains the planet's highest peaks, the seven-mile-high Maxwell Montes. The lowest spot, approximately one and one-half miles below the reference level, lies inside Atalanta Planitia, the giant basin at upper right. Beta Regio, seen at far left, may be a volcanic region akin to the Hawaiian Islands.

A Mercator map like the one at left tends to exaggerate features in a planet's polar regions, a problem that can be avoided by projecting the same information onto a sphere (below). Venus's north pole appears blank because the Pioneer Orbiter did not fly directly over that area.

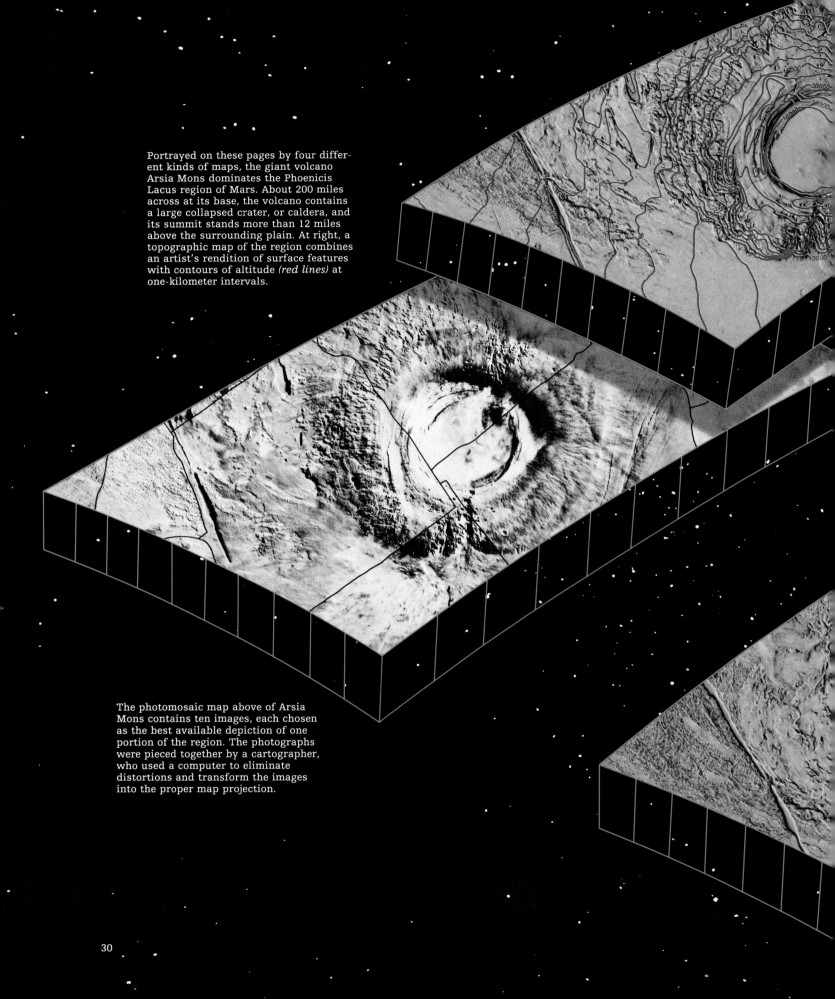

Portrayed on these pages by four different kinds of maps, the giant volcano Arsia Mons dominates the Phoenicis Lacus region of Mars. About 200 miles across at its base, the volcano contains a large collapsed crater, or caldera, and its summit stands more than 12 miles above the surrounding plain. At right, a topographic map of the region combines an artist's rendition of surface features with contours of altitude *(red lines)* at one-kilometer intervals.

The photomosaic map above of Arsia Mons contains ten images, each chosen as the best available depiction of one portion of the region. The photographs were pieced together by a cartographer, who used a computer to eliminate distortions and transform the images into the proper map projection.

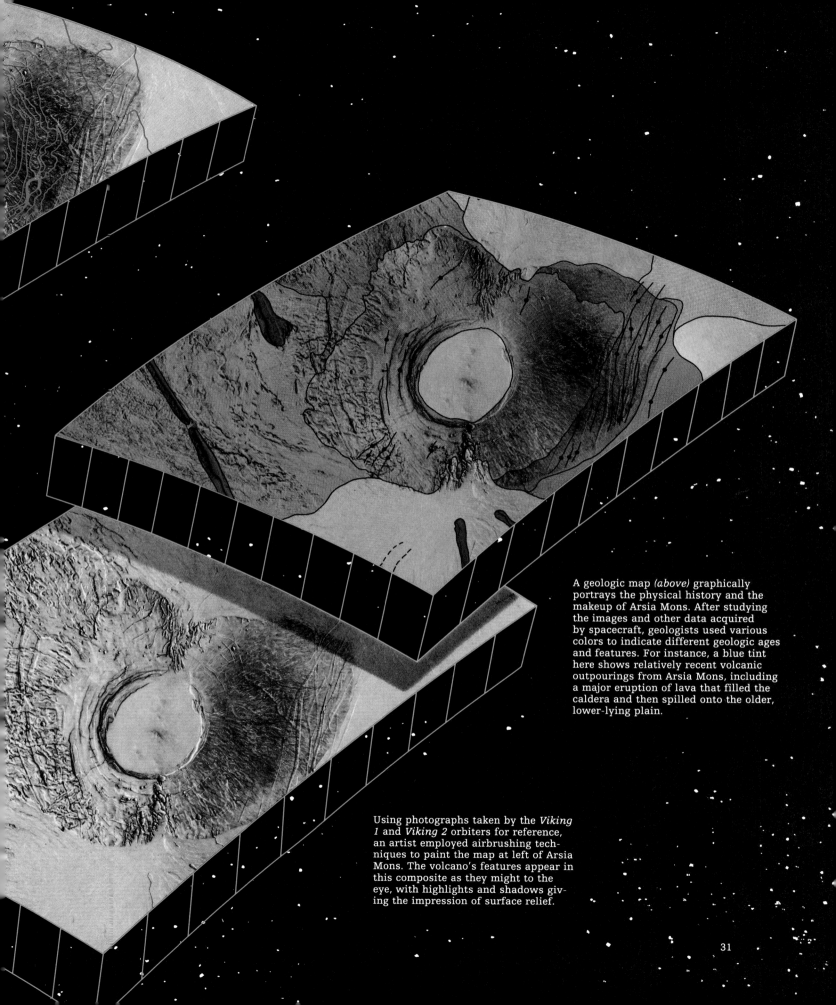

A geologic map *(above)* graphically portrays the physical history and the makeup of Arsia Mons. After studying the images and other data acquired by spacecraft, geologists used various colors to indicate different geologic ages and features. For instance, a blue tint here shows relatively recent volcanic outpourings from Arsia Mons, including a major eruption of lava that filled the caldera and then spilled onto the older, lower-lying plain.

Using photographs taken by the *Viking 1* and *Viking 2* orbiters for reference, an artist employed airbrushing techniques to paint the map at left of Arsia Mons. The volcano's features appear in this composite as they might to the eye, with highlights and shadows giving the impression of surface relief.

water molecules absorb infrared light, astronomers began to look for evidence of water vapor in that portion of the Venusian spectrum. In 1932, American astronomers Walter Adams and Theodore Dunham, using specially sensitized film, took a four-hour exposure of the infrared region of Venusian light and identified its primary absorption lines as those of carbon dioxide, putting the upper limit for oxygen and water vapor at no more than about two percent of their earthly abundance.

There were, of course, alternative views. In 1955, the distinguished British astronomer Sir Fred Hoyle, an indefatigable proponent of radical theories, offered a hypothesis based on possible chemical reactions in a carbon dioxide atmosphere. He proposed that oceans on Venus would not be ordinary, everyday bodies of water but rather oceans of oil. Claimed Hoyle, "Venus is probably endowed beyond the dreams of the richest Texas oil-king." Others, such as American astronomers Fred Whipple and Donald Menzel, held out for more conventional seas. Whipple and Menzel of Harvard University suggested in 1954 that a carbon dioxide atmosphere was inconsistent with the presence of continental landmasses, since the rocks would eventually absorb the CO_2, as they do on Earth. Venus was therefore covered with a planetwide ocean rich in bubbles of carbon dioxide. In other words, Venus was bathed in soda water.

Swampworld, Oilworld, Seltzerworld . . . Venus, disappointingly, turned out to be none of these. A more accurate picture of Venus began to emerge in the late 1950s. No longer the province of solitary astronomers on isolated hilltops, planetary astronomy was yielding to the sustained assault of international scientific teams employing sophisticated modern instruments. In 1957, ultraviolet photographs revealed dim markings on Venus's clouds. By tracking the markings, scientists learned that the clouds spin around the planet every four days from east to west, the opposite direction from winds on Earth. Radar came into play about the same time, and by 1964, it had revealed an astounding fact. By bouncing radar beams off the surface of Venus and measuring their rate of return, astronomers in the Soviet Union and the United States were able to detect the planet's rate of rotation from distortions in the return signal. They learned that under its swift-passing clouds, Venus itself rotates only once every 240 days or so. Not only does it spin very slowly on its axis, but the motion is retrograde, or clockwise, unlike every other planet except Uranus and Pluto.

Observers also began to study the radio waves that emanate from the body of the planet. The frequency of the emissions suggested to some theorists that the surface of Venus might be extremely hot. By 1968, scientists had upped their estimates of the surface temperature from 570 degrees Fahrenheit during 1956 to as

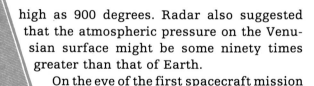

Sketches of Mercury as observed on November 5, 6, 8, and 9, 1882.

high as 900 degrees. Radar also suggested that the atmospheric pressure on the Venusian surface might be some ninety times greater than that of Earth.

On the eve of the first spacecraft mission to Venus in 1962, many uncertainties remained. No human eye had ever seen the planet's inscrutable surface, but the available data now strongly suggested that Earth's twin world was, at best, fraternal: far hotter and deficient in both water vapor and oxygen. The legendary swamps of Earth's sister began to look as illusory as Lowell's Martian canals.

MERCURY THE ENIGMA

If Mars and Venus lent themselves to myths about canals and swamps, tiny Mercury, hugging its mother Sun, resisted all fanciful description. Aided by an early telescope, astronomer Giovanni Zupus discovered the phases of Mercury in 1639, providing further confirmation of the Copernican theory. Three hundred years later, scientists had added virtually nothing to human knowledge of the Solar System's innermost planet. Even with a modern telescope, Mercury is an extremely difficult target. With an average orbital distance of about 36 million miles (versus 93 million for the Earth), Mercury never gets more than 28 degrees away from the Sun as viewed from Earth. To see anything without the Sun's glare, observations must be made near dawn or dusk, but those are also times when the distortions produced by Earth's atmosphere are at their worst. In any case, the planet is so tiny that it remains virtually featureless in even the best ground-based photographs.

Because they could see so little of the planet itself, early astronomers concentrated their efforts on observing the transit of Mercury across the face of the Sun. Then, at least, they could measure the body's position and orbit. Johannes Kepler was particularly interested in these transits because he wanted confirmation that the planet's orbit was elliptical, as his theory predicted. Mercury's transits are rare, however, occurring a mere thirteen times each century, in the months of May and November when the orbits of Earth and Mercury are lined up properly. In May of 1607, Kepler thought he observed such a transit and was so excited that he ran all the way to the castle of his patron, Emperor Rudolf II, to inform him of the event. Long afterward, however, Kepler realized that what he had actually seen was a sunspot—a phenomenon whose existence was not known until Galileo identified it several years later. ("Did I pass off a spot I saw as Mercury?" Kepler asked in 1617. "Then lucky me, the first in this century to observe sunspots.") Although he then predicted an actual transit of Mercury for 1631, Kepler died a year before it happened and his orbital theories were proved correct.

With the orbit known and telescopes improving, scientists in the years after Kepler's death turned to scrutinizing Mercury's surface in search of mean-

Hemispheric map of Mercury, compiled during the years 1924 to 1929.

ingful features. In the late eighteenth century, Johann Schröter reported some "dusky markings" on the planet, but details were absent. By the late nineteenth and early twentieth centuries, several astronomers had noted indistinct surface markings. Among these observers were Schiaparelli and Lowell, both of whom sketched maps showing marks that were (of course) long and linear, not unlike the features they thought they had seen on both Venus and Mars. Later astronomers saw streaks that were less canal-like but still disturbingly vague.

Because the few features that did show up on the surface of Mercury seemed never to change, astronomers concluded that they were seeing only one side of the planet. Some astronomers assumed that this meant Mercury's rotation period was twenty-four hours, the same as Earth's, but Schiaparelli drew a different conclusion. To him, the evidence indicated that Mercury was locked into a synchronous rotation; its day was apparently exactly as long as its year, eighty-eight days. Like the Moon, which always presents the same face to the Earth, Mercury seemed to have one side that always faced the Sun and an opposite side that always looked away.

This theory gained widespread acceptance. Mercury's Sun-facing hemisphere, forever bathed in radiation from the nearby star, was thought to be the hottest place in the Solar System, with a surface temperature of some 700 degrees Fahrenheit. The opposite hemisphere would be a land of eternal night, colder even than distant Pluto. As late as 1960, one astronomer could write with a high degree of confidence: "It is not, therefore, correct to say simply that Mercury is the hottest of the planets. It is also the coldest."

Because no one had ever seen an appreciable atmosphere on the planet, heat from one hemisphere would have no way of traveling to the other side. Astronomers from the University of Michigan were surprised, therefore, when in 1962 their measurements of radio waves from the planet showed that the night side was considerably warmer than predicted. The reason emerged in 1965. In that year, American astronomers Gordon Pettengill and Rolf Dyce at the huge Arecibo radio observatory in Puerto Rico studied the planet with radar techniques similar to those used for Venus. The experiment revealed that Mercury actually rotates on its axis once every 58.6 days. Rather than a one-to-one relationship between day and year, Mercury actually rolls through three days every two Mercurian years. Each side of the planet does periodically face the Sun, so there are no hemispheres of endless day or night. Pluto is in fact colder than Mercury, and Venus is hotter.

By then, Earth-based study of the planets was a fading field, even if it continued to offer up the occasional morsel of knowledge. Although radar, radio telescopes, and spectroscopy were powerful additions to the astronomical arsenal, distance and the obscuring veil of Earth's own atmosphere limited what could be seen of planetary neighbors. But a new and more mobile age was about to dawn. In the following decades, scientists would learn more about the Solar System than they had in the previous three centuries. The best way to learn about the planets was, quite simply, to go to them.

WORLDS IN MOTION

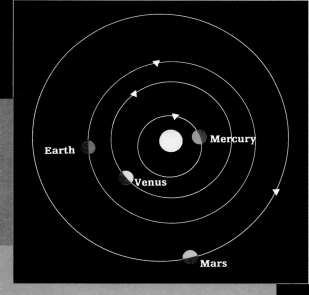

Certainly one of the most far-reaching discoveries in scientific history was the sixteenth-century realization that all planets in the Solar System, including Earth, orbit the Sun. Without an accurate understanding of celestial mechanics—the movement of planets—there would be no space age.

The knowledge came slowly at first. Deriving the true orbit of just one planet, Mars, was the chief toil of German astronomer Johannes Kepler from 1600 to 1608. In the process, he uncovered three general mathematical principles governing planetary motion, and modern observations show that Kepler's laws hold true even for Uranus, Neptune, and Pluto, worlds unknown in Kepler's time: One, a planet's orbit is elliptical, with the Sun at one focus of the ellipse. Among the inner planets *(above)*, Venus and Earth have nearly circular orbits, Mercury and Mars somewhat more eccentric ones. Two, a planet moves fastest near perihelion, its closest approach to the Sun *(triangular marks, above)* and slowest at aphelion, its farthest point. Three, planets nearer the Sun travel faster than those farther out.

Today, Kepler's laws and other principles of celestial mechanics—some of which are described on the following pages—are vital to astronauts and astronomers alike. For example, spacecraft launches and good viewing conditions from Earth are linked to a planet's closest approach *(pages 38-39)*, and safe landings on a distant world require knowledge of its seasonal cycle *(pages 40-41)*.

The Faces of Venus

In ancient times, astronomers thought Venus was two planets, one that appeared in the morning during part of the year, the other seen in the evening during subsequent months. The misconception arose because Venus is a so-called inferior planet, circling the Sun on a path that lies inside of Earth's own orbit. Thus, the Venus *(right)* observed through a telescope waxes and wanes like the Moon and sometimes vanishes altogether. (By contrast, superior planets such as Mars and Jupiter, orbiting farther from the Sun than does Earth, have phases that are much less pronounced.) The phases of Venus follow a 584-day cycle. Twice in each period, the planet disappears from the night sky, aligning with Earth and the Sun in events that are called conjunctions.

An inferior conjunction occurs as Venus passes between Earth and Sun and travels with the Sun through the daytime. Soon after an inferior conjunction, Venus's orbit brings it to one side of the Sun in the sky, a divergence measured in degrees of arc and known as elongation *(below)*. As western elongation grows, Venus becomes the morning star, rising ever earlier before the Sun until a maximum elongation is reached. Then the separation between Sun and planet begins to shrink until Venus vanishes again, this time moving directly behind the Sun in a superior conjunction. Eventually the planet reappears east of the Sun, becoming visible after sunset as Earth's evening star.

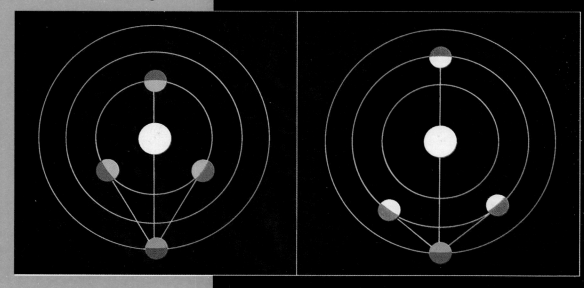

The innermost of the four terrestrial planets, Mercury *(gray)* never travels more than twenty-eight degrees to either side of the Sun *(right)*. When such a maximum elongation is to the west, Mercury may be seen for one hour before sunrise; when to the east, it may be seen an hour after sunset.

Orbiting 30 million miles beyond Mercury, Venus *(far right)* has a maximum elongation that diverges forty-eight degrees from the Sun, allowing a three-hour predawn or postsunset viewing.

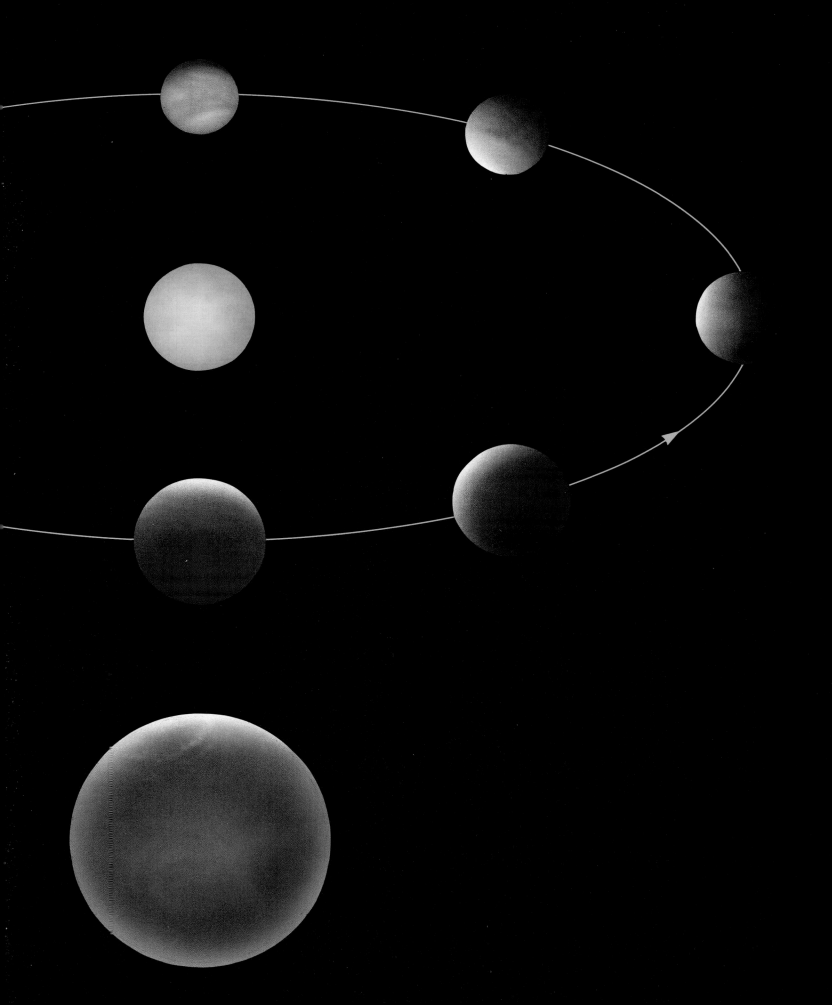

Celestial Flybys

Astronauts planning a voyage to Mars or astronomers trying to reserve scarce telescope time share an interest in the planet's next opposition. That event marks the moment, every 780 days, when Earth lines up between Mars and the Sun, thus putting the Red Planet 180 degrees opposite the Sun's position in the sky. Mars comes nearest to Earth at these times, but because the Martian orbit is more elliptical than Earth's, the actual distance between the two planets during any given opposition may vary by as much as 30 million miles.

As shown in the illustration at right, February oppositions are the poorest, whether for purposes of observation or travel: Mars is then at aphelion, far from the Sun and from the Earth as well. The first of a pair of "good" oppositions—close ones—comes near August every fifteen years (the second happens two years later), when Mars approaches perihelion. Two fairly good oppositions occurred in 1971 and 1973, the period when the *Mariner 9* probe traveled to Mars. Another pair, in 1986 and 1988, was capped by the first Soviet launches to the Martian moon Phobos.

The next good launch and observational opportunity, in the year 2003, will involve the rarity known as a perfect opposition. This phenomenon occurs exactly at the point of closest orbital approach, when the combination of the orbital movements of Earth and Mars reduces the distance between them to a mere 35 million miles. Perfect oppositions occur only once every 284 years. The last one predated the American Revolution by almost six decades.

The average distance between Mars and Earth is 48 million miles at opposition, when the two planets line up on the same side of the Sun. Mars is farthest from Earth— more than 200 million miles—at conjunction, when it lies on the far side of the Sun. Between these two extremes come maximum elongations, when Mars swings ninety degrees east or west of the Sun.

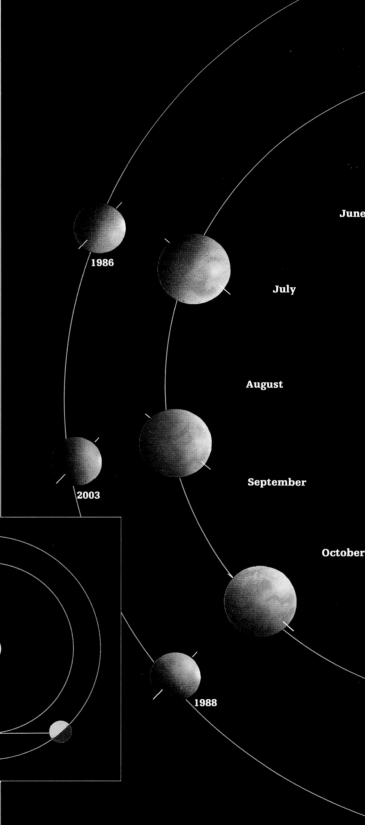

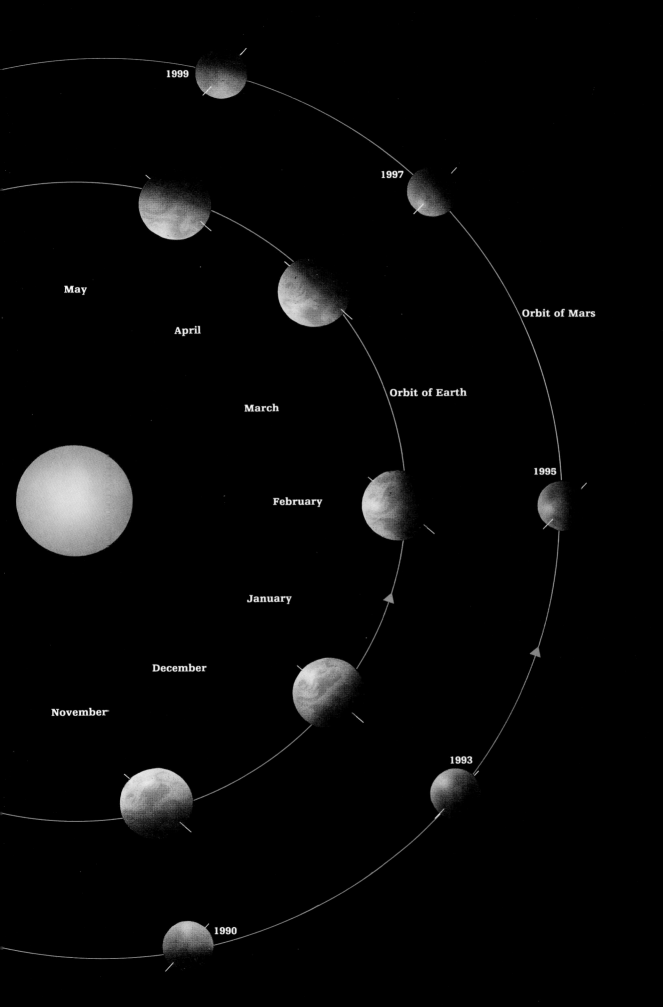

SEASONAL EXTREMES

Tilted to the plane of its orbit by about twenty-four degrees, Mars experiences seasons much as Earth does. Several factors tend to create uniquely Martian weather patterns, however, and also make the seasonal changes, particularly in the southern hemisphere, much more extreme than their terrestrial counterparts. For one, Earth possesses a thick, insulating atmosphere and oceans that store heat. Mars has no oceans and only the most tenuous atmospheric blanket. At low latitudes in the southern hemisphere, the seasonal variation in temperature helps trigger violent dust storms *(pages 114-117)* that mark the Martian southern spring.

The seasonal cycle is also influenced by the planet's eccentric orbit *(below)*, which brings Mars 26 million miles closer to the Sun at perihelion than it is at aphelion. (Earth, in contrast, is only three million miles closer on its closest approach.) Southern summer occurs when Mars is at perihelion, and temperatures in that hemisphere can reach 70 degrees Fahrenheit. This is a 260-degree jump from the minus 190 degrees of the southern winter, which occurs at aphelion. And since the planet is traveling more slowly at aphelion, this savage season lasts six Earth months.

The north, for its part, experiences milder seasons. Although northern temperatures in winter are comparable to those in the south, the northern winter occurs at perihelion, when Mars is orbiting faster. It is thus a comparatively short season that lasts five Earth months. By the same token, the northern summer, at aphelion, is long and relatively cool.

WHEN A DAY IS NOT A DAY

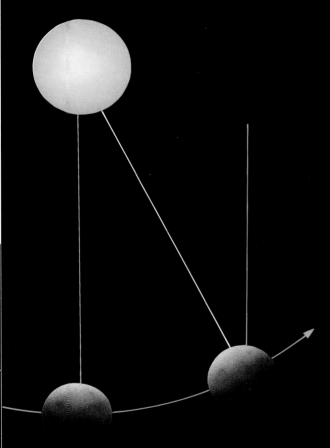

As on Earth, a day on any planet may be defined as the time from noon to noon. But such periods are only "apparent days," sometimes called sols, or solar days, because they are based on the apparent movement of the Sun in the sky. Another astronomical measure is the time needed for a planet to spin once on its axis against the vastly more distant—and therefore seemingly fixed—stars, a period known as the planet's sidereal day, from the Latin for "star."

The difference arises because a planet simultaneously orbits the Sun as it rotates on its axis. Thus, as shown at left, an observer on Mars, measuring from noon to noon, would find that because of the planet's movement along its orbital path *(blue arrow)*, a rotation of more than 360 degrees is needed to bring the Sun once again directly overhead *(white line)*. In contrast, one 360-degree spin is all that is needed between successive appearances of a star overhead.

The Martian sidereal day thus runs twenty-four hours, thirty-seven and a half minutes, but its apparent day runs about twenty-four hours and thirty-nine minutes, a difference of roughly ninety seconds. In the case of Mercury, the difference is considerably more dramatic. Because its rotation is very slow *(page 42)*, and its orbit very fast, the planet's sidereal day is more than 1,400 hours long—but its apparent day is three times longer *(pages 50-51)*.

A Planet
Slowed by Gravity

When the Solar System and Mercury were young, the planet might have completed one rotation on its axis in as little as eight hours. Today a single spin takes about 1,407 hours, or more than fifty-eight Earth days, a slowing brought on by Sun-induced tides. Tidal slowing affects every planet in the Solar System, but none so dramatically as Mercury.

Earthly tides are a complex product of the gravitational attraction of both the Sun and the Moon. The tides of moonless Mercury, in contrast, are subject only to the Sun and, in the absence of oceans, act on the body of the planet itself. Mercury's tidal bulges, greatly exaggerated at right and below for clarity, are very subtle deformations in the planet's rock, only a few centimeters high.

As a planet rotates on its axis and travels in its orbit, the direction of the Sun's gravitational attraction is constantly shifting. Because tides—whether oceanic or solid body—are slow to adjust to these shifts, tidal bulges are always somewhat ahead of or behind the line of gravitational attraction between the two bodies. As Mercury's rotation carries the still-gathering bulge forward, or counterclockwise *(arrow)*, the Sun's gravity tugs it back, a persistent drag effect that has gradually lengthened the planet's day.

Tides come in pairs *(right)* because the Sun's gravity pulls on a planet's particles according to their distance from the Sun. Those on the near side feel the biggest pull *(long arrow)*, creating one of the tidal bulges. Particles on the far side are least attracted: In effect, the rest of the planet moves away, leaving them behind to form a second bulge.

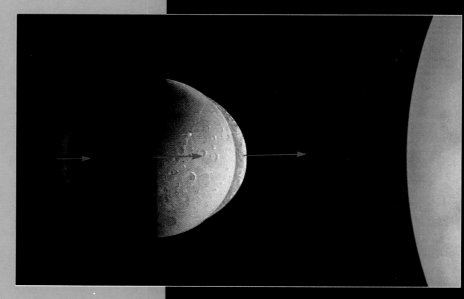

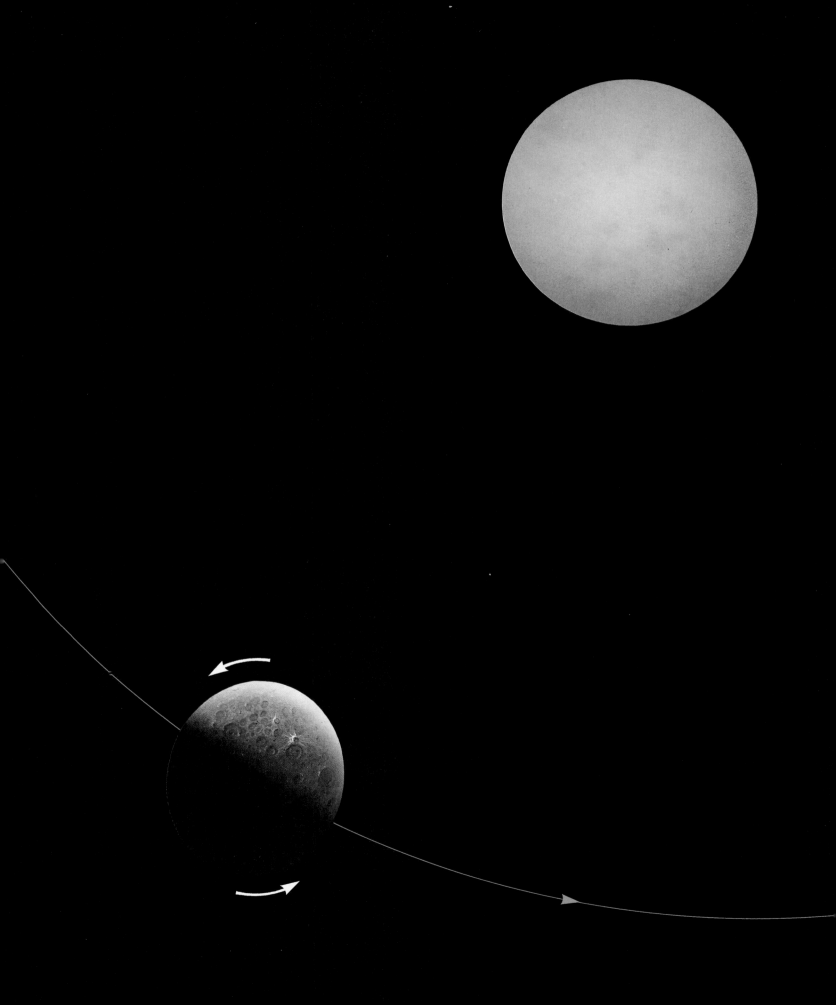

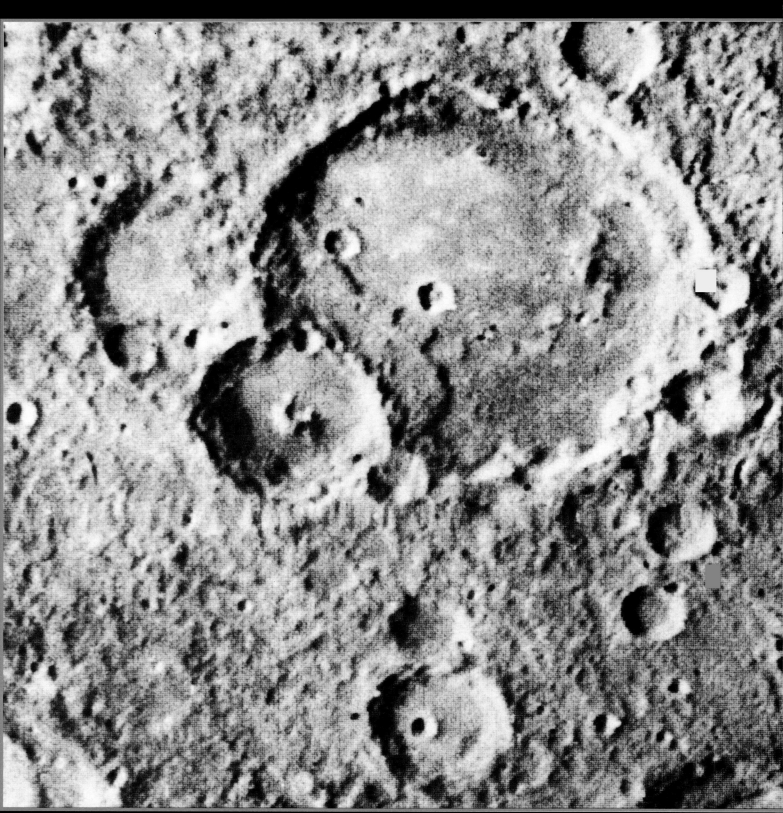

Hidden from Earth by the Sun's glare, Mercury revealed this barren, battered surface to *Mariner 10.* The violent birth of the Solar System is recorded in the planet's ancient basins and their overlying pox of impact craters.

While the Earth was spinning through eighty-eight days, the ancient mile-high peaks of Mercury's Caloris Mountains lay hidden in darkness. But now the eastern ramparts glow in the rays of a slowly rising Sun. To the north, a network of ridges and valleys emerges, a jumble of jagged surfaces and shadowed depths. Mercury's long, frigid night is ending, but the inky blackness of the sky is undiluted by the dawn. Even at noon, it will remain dark.

Nearly four Earth days pass before the Sun slips free of the eastern horizon, its disk looming twice as large as when seen from Earth. Forty days after it appeared, the blazing orb is nearly overhead. Then its progress stops and reverses; for six days, the Sun heads back toward the east before resuming its westward path. The surface temperature, a mild eighty degrees Fahrenheit at midmorning, soars under the protracted high noon, reaching nearly 800 degrees by early afternoon—hot enough to melt zinc. As the Sun descends, the surface begins to cool. By nightfall, eighty-eight days after sunrise, the temperature has plummeted to minus nine.

The disappearance of the Sun marks the end of one of Mercury's years: In the same eighty-eight-day span since first light in the Caloris Mountains, Mercury has come full circle in its orbit. Before the mountains see daybreak again, the planet will complete another lap around the Sun. During the long night, the surface temperature will plunge to minus 300 degrees Fahrenheit— one extreme of an 1,100-degree range of temperature that is greater than that on any other planet in the Solar System.

MYSTERIES OF A DISTANT PLANET

Even so brief a description of a day on Mercury conveys more than scientists knew about the planet two decades ago. Radar observations in the 1960s had established its period of rotation, but most of its other aspects were at best roughly understood. By observing Mercury's gravitational effects on Venus and passing comets, scientists had concluded that the tiny planet, about 30 percent larger than Earth's Moon, was about as dense as Earth. Measurement of Mercury's reflected light revealed a barren, Moon-like surface.

Scientists theorized that, like the Moon, Mercury was pocked by meteorite impact craters and lacked any appreciable atmosphere, but they were less certain about what lay beneath the surface. The planet's surprising density could be accounted for only if it had a metallic core about three quarters its entire diameter. The presumed iron core of the Moon, by contrast, makes up only 20 percent of its diameter, and even Earth's is just 54 percent. Planetary scientists suspected that Mercury's core held clues to the evolutionary process of all the terrestrial planets. Furthermore, they believed its unweathered surface to be an archive of the Solar System's early history. The only way to learn more about Mercury, however, was to dispatch a robot visitor loaded with cameras and electronic instruments.

MOUNTING AN EXPEDITION

At the dawn of the space age, a mission to Mercury was little more than a gleam in the eyes of planetary scientists. One substantial obstacle was cost: With a program that soon included an effort to put an American on the Moon, NASA had doubts about spending millions of dollars on a massive Titan rocket to hurl a half-ton craft across many millions of miles of space to Mercury. It was clear, however, that a smaller, more economical booster would be unable to put an instrument payload on a direct route to the distant planet.

A cheaper course to Mercury emerged in 1962 from research by NASA's planetary exploration center, the Jet Propulsion Laboratory, in Pasadena, California. With the help of computers to compare possible flight trajectories in relation to planetary orbits, JPL staffers found that Earth, Venus, and Mercury would be in a particularly beneficial configuration for a few weeks in 1970 and again in 1973. During one of these favorable launch periods, a spacecraft could be fired into an elliptical solar orbit that would take it through the gravitational field of Venus. Slowed slightly by Venus, the probe would fall into a smaller orbit around the Sun, allowing it to cross Mercury's orbit for an encounter with the planet. By using this so-called gravity assist, NASA could scale back to a smaller booster, the Atlas, and still launch a probe as heavy as a Mariner vehicle, a craft that had already been sent to Venus in 1962. Other Mariners were slated to visit Mars in 1964 and Venus a few years later, in 1967.

The JPL engineers were eager to try a

From 240,000 miles away, *Mariner 10*'s cameras recorded a pastiche of basins and craters on the planet's sunlit hemisphere hours before the probe's first close approach.

gravity-assisted mission to Venus and Mercury not only for the immediate reward of visiting Earth's nearer neighbors but also to gain experience for more venturesome missions to Jupiter and the outer Solar System. As it happened, however, budget-conscious NASA administrators delayed making a decision. The project did not get the go-ahead until 1969. The years needed to prepare the exploratory vehicle precluded liftoff in the 1970 launch period. That left the thirty-seven-day span between October 16 and November 21, 1973, the only other suitable interval before the 1980s. JPL would have four years to get ready.

Though tight, the deadline was manageable. The budget would have to be closely watched, however. NASA officials declared that the Venus-Mercury probe would be the standard-bearer for a new line of low-cost, high-efficiency missions. In December 1969, JPL director William Pickering committed his organization to a fixed price of $98 million for the entire project. Pickering intended to economize by using a modified Mariner space vehicle—by now proven in flights past Mars and Venus—and hardware left over from earlier

In a bit of interplanetary thievery, *Mariner 10* stole energy from the gravity and orbital motion of Venus *(below, left)* to reduce its speed relative to the Sun. This bent its trajectory into a solar orbit that brought it past Mercury *(below)* at six-month intervals. The first space probe to use such a gravity assist, *Mariner 10* had to be aimed with great precision: Missing a 248-mile-wide portal 10,000 miles from Venus would have sent the craft careering through space, forever forfeiting its date with Mercury.

missions. Walker E. (Gene) Giberson, a veteran of JPL's Surveyor Moon-probe program, was Pickering's choice to manage the project. Known among his colleagues as an incurable optimist, Giberson also had a reputation for overcoming obstacles to get the job done.

NASA convened the Science Steering Group to help plan the mission, which would later be known as *Mariner 10.* This marked a break with the early years of planetary exploration, when the challenge of getting to a target planet was so great that the design of a spacecraft and its subsystems was left almost entirely to engineers; the payload of instruments was usually taken up late in the process. This time, scientists representing the disciplines that would be involved in the closeup study of Mercury sat down with engineers to shape the project from the beginning. They established scientific priorities for *Mariner 10,* deciding what phenomena should be studied and recommending suitable experiments. In addition to cameras, the spacecraft would carry instruments to provide information for calculating Mercury's mass, diameter, and surface temperature and to look for signs of an atmosphere. Other devices would examine the planet's interactions with the solar wind, a tenuous flow of charged particles pouring from the Sun.

Each of *Mariner 10'*s six experiments would be the responsibility of a principal investigator, who would develop the data-gathering device and perform the first analysis of information as it streamed back from the spacecraft. Supervising this diverse group was James Dunne, an earth scientist who had served on previous lunar and planetary exploration teams. The television imaging system that would take pictures of the planets was entrusted to a team of scientists headed by geologist Bruce Murray, another old hand; he had worked on four previous interplanetary flights.

Shortly after the Science Steering Group was formed, the prospects for scientific yield were greatly improved by the chance intervention of Giuseppe Colombo of the Institute of Applied Mechanics in Padua, Italy. Several years earlier, this prominent astrophysicist had explained the unusual relationship between Mercury's spin and its orbital period *(pages 50-51).* In early 1970, after attending a presentation of plans for the upcoming mission at the California Institute of Technology, JPL's parent institution, Colombo buttonholed Bruce Murray and asked a startling question. "What will be the period of the spacecraft about the Sun after the Mercury encounter? Can the spacecraft be made to come back?" "Come back?" Murray asked. "Yes," Colombo replied, "the spacecraft could return to Mercury."

Murray asked a JPL navigation engineer to investigate the notion. Soon he had his answer: Among the possible courses for *Mariner 10* was one that would, with minor in-flight corrections, mesh its orbit with Mercury's so that the planet and the probe would meet every other time Mercury came around the Sun. The "Colombo connection," as it was called, dictated changes in the basic Mariner spacecraft, and JPL engineers soon had a growing list of modifications. Because the probe would pass closer to the Sun than any previous spacecraft, its equipment would be shielded with special insulative

Two-Year Days and Other Anomalies

In accordance with Kepler's third law of planetary motion *(page 35)*, Mercury, as nearest to the Sun, also travels fastest: It completes one orbit, or one Mercury year, every eighty-eight Earth days. Yet the planet spins so slowly on its axis that in two orbits it completes only three sidereal days, or full rotations against the background stars *(page 41)*. An apparent day on Mercury, as perceived by an observer on the planet, takes three times longer *(below)*. This three-to-two relationship—or 3:2 spin-orbit coupling, as it is known—is unique in the Solar System. The Moon's one-to-one spin-orbit relationship is more common.

A complex of factors influence the unusual length of a Mercury day with respect to that of its year. One is the tidal slowing that accounts for the planet's tortoiselike spin rate *(pages 42-43)*. Another factor is Mercury's nearness to the Sun at perihelion, when the Sun's gravitational pull on the planet is at its strongest. Mercury's orbital motion speeds up so much at perihelion that its already slow rotational motion cannot, in effect, keep up. At some locations on the tiny globe, the Sun actually seems to backtrack in the sky at noon. In other spots, the result of the anomalous spin-orbit relationship is a double sunrise *(opposite)*.

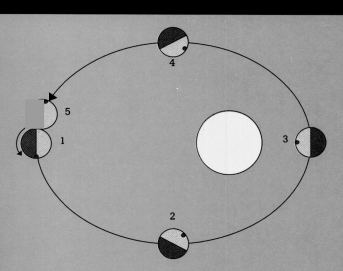 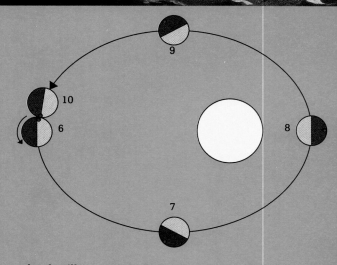

As shown in the views looking down on Mercury's orbit, an observer *(red dot)* on the planet would experience a day that lasts two Mercury years. If sunrise, when the observer is carried out of Mercury's night side, takes place when the planet is at aphelion *(1)*, then noon, when the Sun is directly overhead, will occur at perihelion *(3)*; sunset will fall at the next aphelion *(5)*. A second year begins *(6)*, and the observer orbits counterclockwise through night until the next aphelion, when sunrise—and the start of a third year—marks the completion of one Mercury day *(10)*.

An observer *(red dot)* views an unusual double sunrise during Mercury's closest approach to the Sun. As the planet's counterclockwise spin moves the observer out of darkness, the Sun rises for the first time *(1)*. Nearing perihelion *(2-4)*, the planet accelerates until its orbital speed matches and then exceeds its spin rate. This has the effect of rotating the observer back into darkness *(5)*, and the Sun sinks below the horizon *(above)*. As Mercury moves past perihelion *(6-8)*, its orbital speed drops below the spin rate; the observer again crosses the day-night line *(9)*, and the Sun rises for the second time.

fabric and sunshades. Winglike solar panels, which would convert sunlight into electricity, had to be mounted on pivots so they could be tilted to avoid the full power of the Sun's searing radiation.

One particularly troublesome—yet vital—item on the engineers' list was the telecommunications system. It would carry commands to *Mariner 10* and send flight information and scientific data back to Earth. The computerized messages would be streams of digital signals, called bits, transmitted at either 22 or 117 kilobits per second. (A kilobit is a thousand bits.) The lower rate would suffice for most experiments, but the images taken by Mariner's two high-resolution television cameras demanded greater capacity. A single image would contain more than five million bits of information, and because the on-board tape recorder for data storage had only a limited capacity, *Mariner 10* would have to transmit each image at once or lose many pictures forever. Members of the imaging team worried that when *Mariner 10* reached Mercury, then 90 million miles from Earth, the signal from its twenty-watt transmitter would be too faint for the 117-kilobit rate to be reliable. At the 22-kilobit rate, however, four-fifths of the possible pictures of Mercury would never be returned to Earth.

Project engineers thought up and discarded one remedy after another. A higher-frequency transmitter was vetoed by NASA on the grounds that it might cost as much as two million dollars. The next idea was to enlarge *Mariner 10*'s high-gain antenna, a fifty-four-inch rotatable dish that was already much more powerful than the cone-shaped omnidirectional antenna at the other end of the vehicle. This alternative was also rejected: A larger high-gain antenna would not fit under the aerodynamic housing that would protect the spacecraft during launch. Then project managers reconsidered the possibility of a new transmitter, since JPL's frugality in other areas had squeezed out enough money to pay for it. But the battle of the bits, as it became known, was not yet won. The new transmitter, finished just four months before the scheduled launch, failed to work in its final test. *Mariner 10* would have to fly with its original radio system.

A COOL SOLUTION

One expedient remained. If the spacecraft could not send stronger signals, perhaps the huge, 210-foot receiving antennas of NASA's Deep Space Network could be improved. These steerable dishes, located in Spain, Australia, and California, were already among the most sensitive in the world, but only a few weeks before the launch, Deep Space engineers found a way to make them better. The apparatus that amplified incoming transmissions—normally refrigerated to minus 430 degrees Fahrenheit to reduce heat-generated interference from the equipment itself—was cooled further to minus 434 degrees. The incremental temperature change gave a significant boost to the antennas' reception of very faint signals. After a few adjustments were made at the ground stations and on the spacecraft to take advantage of this new capacity, the TV team was finally happy. For the first time since mission planning began, they were confident of obtaining plenty of high-resolution images of Mercury.

Shortly after midnight on November 3, 1973, four years of toil and hope soared skyward with *Mariner 10* from the Cape Canaveral Air Force Station in Florida. After a programmed two-minute burn, the Atlas booster fell away and the upper-stage Centaur punched the craft into orbit 117 miles up. Nearly a third of the way around the Earth, the Centaur engines fired again, pushing *Mariner 10* into an elliptical path around the Sun at 25,458 miles per hour. Leaving the Centaur behind, the spacecraft unfurled its delicate solar panels and stretched its instrument-bearing arm. The mission had begun.

Fully loaded, *Mariner 10* weighed 1,175 pounds. The fuel tank for its main engine held sixty-four pounds of hydrazine—enough for a total burn time of 550 seconds. Another tank contained eight pounds of compressed nitrogen for the attitude-control system. The gas would be vented through three pairs of jets around the spacecraft to counter its inherent tendency to drift out of position, and to nudge it into the proper orientation for critical functions such as engine firings and planetary photography. Most of the scientific experiments were bolted on the body of the spacecraft; two magnetometers rode on the instrument boom. Perched on a movable platform atop the body were the two television cameras and an ultraviolet spectrometer.

TROUBLE EN ROUTE

With the craft on its way, project engineers at JPL's Space Flight Operations Facility in Pasadena began a vigil in front of computer screens and printers, monitoring telemetry data about every flight system and instrument. They had scarcely settled in their chairs when the first crisis arose. Thermal sensors aboard the spacecraft showed a failure in the heaters designed to keep the television cameras in their optimum temperature range of forty to sixty degrees Fahrenheit. Without the heaters, the extreme cold of deep space could distort the alignment of the telescope optics, causing the cameras to go completely and permanently out of focus and thus blinding the spacecraft long before its rendezvous with Mercury.

Mission controllers tried everything they could think of: They radioed commands to *Mariner 10*'s central computer to turn the heaters off, then on, to no avail. They studied ways to divert heat from another section of the craft to the cameras but were stymied by the insulation designed to protect equipment from the heat of the Sun. Anxious members of the TV team could only wait as the cameras returned their first test pictures of the Earth. Magnifying glasses in hand, they examined the shots. At minus fifteen degrees, the optics seemed unharmed; the images were crisp and even showed small-scale storms sweeping across the planet. But as the cameras grew colder, photos they took of the Moon showed signs of diminished resolving power. Then, on the evening of November 5, the temperature seemed to bottom out at minus twenty-two degrees. Engineers concluded that the heat produced by the operation of the cameras themselves was keeping the temperature from falling further. They decided to leave the cameras switched on, gambling that their delicate electronics could hold up under five months of nonstop operation.

Ten days into the flight, *Mariner 10* gave its controllers another scare during a maneuver designed to refine its trajectory for the gravity-assist encounter with Venus. The operation was the first full-scale workout for the vehicle itself, and at the outset everything seemed to go perfectly. The nitrogen jets of the attitude-control system pushed the probe through a slow pirouette until internal gyroscopes indicated that the main engine was positioned to propel the craft in the right direction. The hydrazine rocket was fired for twenty seconds, just enough to produce the desired velocity change. Then the nitrogen jets rolled the spacecraft back to its original cruise attitude, with its high-gain antenna pointed toward Earth.

Mariner 10 used celestial navigation to maintain this position, relying on two sensors, or trackers, that could determine the direction to the Sun and to Canopus, a bright star visible from Earth's southern hemisphere. Flight controllers breathed a collective sigh of relief at the end of the trajectory correction when the Canopus tracker locked back onto its target. But then the telemetry data showed that the spacecraft had begun to roll, without instructions from Earth. The star sensor had lost contact with Canopus, and the gyroscopes had automatically switched on, guiding *Mariner 10* through slow rolls in an effort to find the star again. For the next ninety minutes, controllers held their breath until the tracker locked back onto Canopus.

Subsequent analysis revealed that the sensor had probably been distracted by a bright dust particle, possibly shaken loose during the rocket burn and now traveling on a parallel course. Mistaking this speck for Canopus, the sensor had tried to align the vehicle with a moving object rather than the fixed star. Unfortunately, there seemed to be no way to avoid the problem. Controllers would simply have to stay alert to prevent excessive loss of attitude-control gas in similar situations.

A week later, on November 21, the spacecraft faltered again. As the navigational gyros were turned on at the start of a routine test, an anomaly occurred in a crucial subsystem that processed scientific data. There had apparently been a sudden surge of power that raised the possibility that it might fail completely at any time. Gun-shy project managers decided to cancel the remainder of the test. On December 7, the same gyros were turned on again while engineers monitored telemetry channels for disturbances in the electronic components. All performed properly except the flight data system,

En route to Mercury, *Mariner 10* employed its solar panels as sails to capture the faint pressure generated by sunlight, using this slight force to control the craft's roll rate. Varying the orientation of the two panels kept *Mariner 10* in a proper cruise attitude with only minimal help from its nitrogen-gas attitude-control thrusters. More energy conserving than propulsive, solar sailing, as the technique is known, allowed the probe to stretch its dwindling nitrogen supply enough to complete two additional flybys of Mercury.

where the same symptoms recurred. JPL technicians, combing vainly through their telemetry records for an explanation, began to wonder if there might be some fundamental flaw in the spacecraft's power system.

The question arose again within three weeks when a more serious malfunction surfaced. On Christmas Day, the strength of the signal from *Mariner 10*'s high-gain antenna plummeted inexplicably. This had little immediate effect, since the probe was still near enough to Earth that even the less powerful signal could carry a full load of data. But communications would have to work perfectly by the time *Mariner 10* reached Mercury in March, or no high-resolution pictures would be returned. Hearing of the latest calamity, TV team leader Murray said, "Sometimes I'm not really sure God is on our side in this mission." Then, after four days of faint signals, the antenna began transmitting at full power. Four hours later, it failed again; five days after that, signal strength returned. Speculating that a key part of the antenna might be too cold, controllers tried to orient the spacecraft so that it would be warmed by the Sun, but after two days the trouble returned. Discouraged engineers began to devise alternatives for image transmission. The worst-case scenario assumed that data would be passed at just over six percent of the hoped-for rate.

By mid-January, *Mariner 10* was three weeks away from its crucial flyby of Venus. Preparations began for a second trajectory correction maneuver to refine the craft's course. Without such a change, the probe would get a gravity assist of the wrong magnitude from Venus and miss Mercury by nearly a million miles. Flight controllers sent their commands to the probe five days in advance; the instructions were stored in the memory of an on-board computer so they could be carried out even if contact was lost at a critical moment. When the time came, however, *Mariner 10* executed the complicated maneuver without a hitch. In early February, after ten days of tracking, the navigation team confirmed that *Mariner 10* would pass Venus within seventeen miles of the aim point.

It seemed for a while as though the capricious spacecraft had changed its ways. Not only had the probe come through the course correction in fine form, but the faulty camera heaters had spontaneously turned back on. The good behavior did not last long, however. On January 28, just eight days from Venus, controllers began a roll maneuver to calibrate the magnetometers by pointing them in all possible roll directions. Suddenly a telemetry alarm flashed on a screen at mission control, warning that orientation gas was being expended at a rapid rate. The attitude-control system had suddenly gone haywire, causing violent oscillations that ended only when the gyros were turned off. By then, however, more than a pound of nitrogen had been lost. With the most important part of the mission still ahead, nearly half of the original eight pounds of attitude-control gas was gone.

Mission controllers were reluctant to depend on the suspect gyros during the Venus flyby, when attitude control would be critical to keeping instruments and cameras pointed at their targets. Casting about for an alternative, they decided to use the star trackers instead. It was a risky choice, since the brightness of the planet itself might distract the Canopus tracker; in that event, the spacecraft would return pictures of empty space.

On the morning of February 5, images of Venus began to arrive in Pasadena. At 10:00 a.m., *Mariner 10* made its closest approach, within 3,600 miles of Venus's cloud tops, still sending back pictures of the planet and still firmly locked on Canopus, despite the brightness of Venus, now three-quarters illuminated by the Sun. Eight days and 4,165 images later, the portrait session came to an end; *Mariner 10*'s cameras and instruments had given Earth-based astronomers a wealth of new information about Venus's atmosphere, ionosphere, and interaction with the solar wind. Then, with Venus's gravitational help, the spacecraft was on its way toward Mercury.

SIX WEEKS AND COUNTING

Though jubilant over the success of the Venus encounter, the ground team still had plenty to worry about. Mercury was six weeks away, and the spacecraft remained plagued by troubles. During the trip to Venus, the electrical system had spontaneously and irreversibly switched to its back-up circuits; if these failed, the mission would end. The high-gain antenna was transmitting weak signals, the oscillation problem in the navigation system persisted, and the level of attitude-control gas was dangerously low.

Then on February 14, in a test of the gyros preparatory to a third trajectory correction, the system ran wild again, wasting more nitrogen. Project managers canceled the correction maneuver as too risky, deciding to wait a month, until mid-March, when the spacecraft's ordinary cruise orientation would point its main engine directly toward the Sun. A rocket burn at that time would refine the probe's orbit so that it would intersect Mercury's. The so-called Sun-line maneuver would avoid complicated rolling and pitching, but the decision introduced another peril: Because the course change would come very late in the voyage, there would be no opportunity for another correction if it proved unsuccessful. The Sun-line maneuver would be *Mariner 10's* last chance to get on the best path to Mercury.

The wait was fraught with further attitude-control problems. By early March, the spacecraft had wasted so much orientation gas that increasingly desperate project managers decided on a radical effort to salvage the mission: They would turn *Mariner 10* into a celestial sailboat. First they overrode the automatic stabilization system; henceforth the gyros would be activated only on instruction from mission control. Then they trimmed *Mariner 10's* solar panels to catch the tiny force of radiation from the Sun. With slight adjustments of the panels and economical bursts of nitrogen, flight directors found they could control the spacecraft's roll rate. By March 12, they had stabilized *Mariner 10* in its normal cruise orientation.

On March 16, the Sun-line maneuver began. The main engine fired for 51.1 seconds, producing a velocity change of forty miles per hour directly away from the Sun. The anxiety pervading mission control was palpable, as engineers and scientists waited for signs that the tiny adjustment had worked. Days later came the good news: *Mariner 10* was on course. It would pass Mercury on its dark side—and only seventeen minutes later than originally scheduled. Meanwhile, another pleasant surprise had occurred. On March 17, the high-gain antenna had spontaneously returned to full power and stayed there. For the moment, at least, *Mariner 10* was on its best behavior.

THE THREE-MILLION-MILE VIEW

On March 24, the probe's cameras began to focus on Mercury. In Pasadena, journalists and scientists crowded the JPL auditorium, intently watching a television monitor that flickered with the first images of the Solar System's innermost world. Still three million miles from the probe, Mercury appeared only as a thin arc of light. But *Mariner 10* was closing the gap at more than 23,000 miles per hour. Soon the earthbound audience began to see a mottled crescent and then the blossoming of individual features on the surface. One of the most distinctive was a crater with bright rays issuing from its center. The first feature observed on Mercury, it was dubbed Crater Kuiper, in honor of Gerard Kuiper, a member of the TV team who had died just six weeks after the spacecraft was launched.

Kuiper, a pioneering planetary scientist and longtime advocate of space exploration, had predicted that Mercury's surface would have both lava fields and impact craters. As *Mariner 10* swooped closer to the planet, TV team members saw evidence that he had been correct. The cameras revealed craters spattered across Mercury's dusty-looking face and vast plains that resembled ancient terrestrial lava flows. In some regions, snaking lines of cliffs two miles high reminded geologists of the earthquake escarpments produced on Earth when one crustal plate is thrust over another.

By March 29, as *Mariner 10* neared its closest approach, members of the TV team were scrambling to interpret the swelling flood of images. In the mean-

THE AFTERMATH OF IMPACT

Airless and stable, Mercury is a ready-made laboratory for the study of crater formation. Because the planet orbits so close to the Sun, it is a prime target for comets, which account for 40 percent of recent cratering of Mercury—as compared to only three percent for Mars. The effects are dramatic: Comets smash into Mercury at nearly fifty-three miles per second—four to six times faster than the asteroids and meteoroids that also bombard the planet—and produce correspondingly outsize craters. (Meteoroids that strike the surface are called meteorites.)

Even a small incoming object can release energy equivalent to the explosion of tens of thousands of hydrogen bombs. According to one theory for the formation of so-called complex craters *(right)*, the enormous force of the strike triggers an interplay of shock waves and ejecta, or material thrown out by impact. The result is a cavity with a well-defined rim, floor, terraced walls, and a central peak. On Mercury, ejecta can form a blanket or skirt around the crater as wide as the diameter of the crater itself. (Ejecta blankets on the Moon, with roughly half Mercury's gravity, are even wider.) Fragments falling farther out create secondary craters and powdery patterns called rays.

A complex crater displays all its features: central peak, flat floor, and terraced walls.

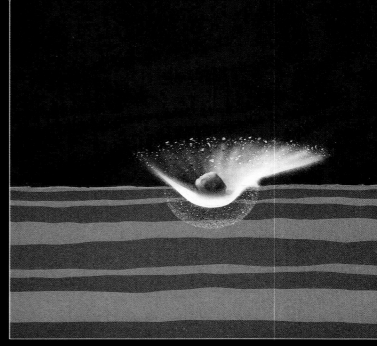

A high-speed meteorite collision sends shock waves *(white arc)* traveling at tens of thousands of miles an hour through Mercury's crust, compressing the impact site and the projectile itself.

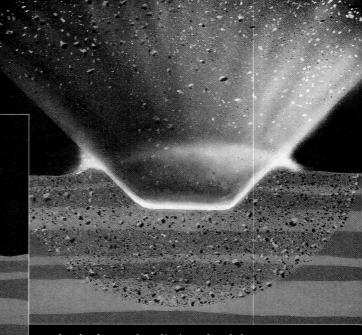

The shock wave has dissipated and the crater stops growing as the force of gravity eventually overcomes the force of ejection.

A fountain of pulverized and melted rock spouts from the growing crater as the crust rebounds, breaking into fragments that escape outward at an angle.

The meteorite rapidly decompresses and explodes: Some of it vaporizes, and some shatters into flying molten fragments. As molten rock lines the crater cavity, the crater grows wider rather than deeper.

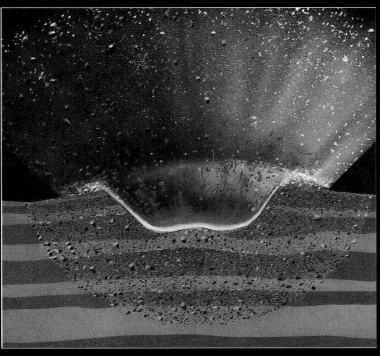

As the pressure above it drops suddenly, a cone of rock shoots up from deep below the impact point. At the same time, the last rocky fragments form a raised edge around the crater.

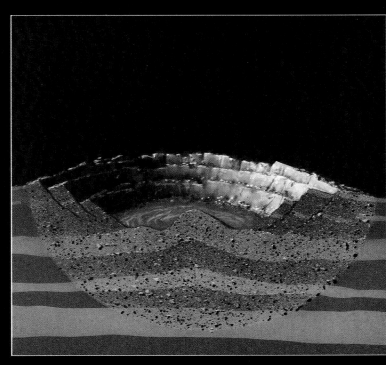

The ascending cone tugs at the crater floor and pulls down the rim, creating a stepped profile as the walls slide inward. The collapsing rim and molten rock further fill in the crater's floor.

MERCURY'S IRON HEART

In the years following the flight of *Mariner 10,* planetary scientists studying its data have put forth a variety of theories about the origin of Mercury's iron core. Each amends the standard model for the accretion of the terrestrial planets. That model explains most of the observed densities of the planets but cannot account for the large size of Mercury's core in comparison to its mantle, the layer of lower-density rock just below the outer crust.

One hypothesis holds that in the maelstrom of the Solar nebula, the aerodynamic characteristics of iron-rich planetesimals caused them to concentrate disproportionately in the innermost regions of the Solar System, where Mercury took shape. When the planets began to coalesce, Mercury would have received more than its share of iron and correspondingly less of the lighter silicates that predominate in the mantle.

In two other models, Mercury is assumed to have formed with a normal amount of iron for its accretion zone, then to have lost a significant portion of silicates early in its history. According to one of these scenarios, the intense radiation and solar wind of the young Sun vaporized and stripped off most of Mercury's silicates, leaving the iron core intact. The other scenario posits a planet-size object striking Mercury and blasting away a large portion of the mantle.

Since each hypothesis predicts a different composition for the mantle, scientists wishing to examine Mercury in the necessary detail will need a new interplanetary probe capable of orbiting the planet.

Another objective of such a mission would be to follow up on the 1985 discovery of sodium and potassium in Mercury's atmosphere. Two scientists in Texas found evidence of these elements by spectroscopic analysis of light from the planet at wavelengths that *Mariner 10* had not examined. The sodium, though very rarefied, appeared to be about five times as dense as the helium detected by *Mariner 10;* the potassium was one-tenth as dense as the helium. Since astronomers believe that the sodium and potassium are torn from surface minerals by both the solar wind and meteoritic bombardment, closer analysis of the atmosphere could yield clues to the presence of other materials in the crust.

time, other researchers watched incoming scientific data. They paid particular attention to three experiments that monitored Mercury's interplay with the solar wind, expecting to see their readings drop to zero as the probe passed into the lee of Mercury. But suddenly, just nineteen minutes from the encounter, the detectors registered a dramatic increase in the concentration of charged particles and a change in their direction of flow. To scientists acquainted with the behavior of the solar wind around Earth, the data suggested a phenomenon called a bow shock wave—the result of interaction between streaming charged particles and a planetary magnetic field.

No one had predicted such a finding. According to prevailing theory, a planetary magnetic field was the product of a rapidly spinning planet with a massive, partially molten metal core; Mercury's leisurely fifty-nine-day rotation seemed far too slow. But *Mariner 10*'s magnetometers buttressed the indications of the other instruments: The level of magnetism was rising as the probe neared the planet. All signs were that Mercury did indeed possess a magnetic field, albeit a weak one.

Other experiments were producing results closer to the investigators' expectations. Measurements of ultraviolet radiation suggested an extremely thin atmosphere, made up mainly of helium with traces of hydrogen. Although atmospheric gases were detectable as much as 620 miles above the planet's surface, the pressure was less than a ten-trillionth of Earth's—closer to a perfect vacuum than anything achieved in terrestrial laboratories. Researchers used infrared readings to estimate the rate at which the planet's surface cooled after dark. From this information they deduced that Mercury, like the Moon, was covered by a loose, porous layer of fine-grained dust. Precise tracking of *Mariner 10* as it moved past the planet told scientists how gravity had affected the probe's trajectory, allowing them to calculate a value for Mercury's mass a hundred times more accurate than any previous measurements. Coupled with the evidence for a magnetic field, this finding supported conjectures that Mercury had a large iron core.

BACK ON THE ROCKY ROAD

Mariner 10 had performed admirably, but before anyone on the ground could grow too complacent, the craft resumed its errant behavior. One day after the closest encounter, a precipitous, unexplained temperature rise in the instrument bays convinced many experimenters that complete failure of the probe was imminent. Some demanded that all efforts be devoted to collecting and transmitting as much data as possible, subverting the long-term program for the mission, already carefully worked out and stored in the spacecraft's computer. But others on the flight team protested, including chief scientist Jim Dunne. Storming into project manager Giberson's office to pound his fist on the desk, Dunne insisted that the flight proceed as planned. He carried the day; the temperature rise was stopped by turning off the cameras and other power-draining instruments.

Soon another debate rose among the scientists. Less urgent but nonetheless important, the issue was how *Mariner 10* should pass Mercury on its second encounter, six months hence. To bring the probe near enough to the planet to make useful observations, a trajectory correction was necessary; uncorrected, *Mariner 10* would pass at a range of 500,000 miles. Many scientists wanted a second dark-side passage, the better to study the newly discovered magnetic field. But the imaging team wanted the second pass to be over the sunlit side to allow photographic coverage of more of the planet. Since a relatively distant passage on the day side would put the probe in a better position to achieve a third encounter, the Science Steering Group, after hours of deliberation, sided with the TV team: *Mariner 10* would return to its target's sunlit face. If a third approach could be squeezed out, it would be devoted to a close dark-side flyby.

This time the course change could not be a Sun-line maneuver. Instead, controllers would have to reorient the probe for two separate rocket burns, each time risking wild oscillations that might exhaust the now extremely limited supply of attitude-control gas. Early in the afternoon of May 9, the spacecraft rolled and pitched into position for a 195-second rocket firing, then successfully twisted back to its cruise orientation, regaining its celestial reference points without incident. A day later, it repeated the performance. During each burn, the spacecraft operated normally, with no vibrations or excessive gas loss. The velocity change was within one percent of the requirement, and it looked as though there might still be enough gas to make a third encounter possible.

In June, communications were temporarily broken when *Mariner 10,* now 148 million miles from Earth, passed on the far side of the Sun. When it emerged from the solar shadow, analysis of its radio signals indicated that although the correcting maneuvers had been successful, a slight trajectory adjustment would be required to make the third encounter possible. The correction would be the trickiest of all, because during the maneuver the spacecraft would have to turn its high-gain antenna away from Earth. With communications broken, *Mariner 10* would rely on instructions stored in its memory, first to make the maneuver and then to turn back to reestablish links with mission control.

The fateful burn was set for July 2. The command sequence was sent to *Mariner 10,* and in the early afternoon, tense controllers saw their telemetry signals go dead as the maneuver began. Their temperamental charge was on its own in deep space. There was no way of knowing whether it was performing normally or oscillating and wasting the last of its precious orientation gas. Finally a faint signal came through to the Deep Space antenna at Goldstone Lake

in the California desert. The burn was concluded, and *Mariner 10* was turning itself back toward Earth. When the pens of the telemetry system began to write again, the engineers read happy news: The spacecraft was on course for both a second and a third pass at Mercury, and no gas had been wasted.

On September 21, after a complete revolution around the Sun, *Mariner 10* flew by Mercury again, at a distance of 30,000 miles. The cameras transmitted several hundred pictures to Goldstone, providing coverage of the previously unseen south polar region and of the southern part of the sunlit hemisphere. As they studied the images, scientists became increasingly convinced that Mercury had known periods of heavy geologic activity. For instance, the gently rolling terrain between heavily cratered regions had probably been formed by flowing sheets of lava; these so-called intercrater plains covered

Ripples frozen in stone commemorate the colossal impact that dug the Caloris Basin—one of the largest craters in the Solar System, with greater area than France. The Caloris Mountains, mile-high chunks of crust thrown up by the impact, mark the basin rim *(blue outer ring).* Within this ancient well, the plain is scarred by concentric ridges and fractures *(blue inner rings),* possibly caused by vertical flexing after the collision.

nearly half of the illuminated face. Planetary geologists suspected that Mercury's other distinctive features, the monumental scarps, had been created by wrinkling of the crust. The scientists estimated that Mercury's most impressive impact basin, 800-mile-diameter Caloris, had been created by an object 100 miles in diameter. Such was the force of this ancient collision that its shock waves had apparently produced jumbled and fractured corrugations—informally labeled the "weird terrain"—on the opposite side of the planet.

PREPARING FOR THE LAST DANCE

As *Mariner 10* left Mercury again, preparations began for the third and final encounter, scheduled for March 1975. As ever, the primary concern was the dwindling supply of attitude-control gas. Even in the euphoria of the successful second pass, project manager Giberson said he expected the next few months "to have some of the elements of 'The Perils of Pauline.' "

He was right. On October 6, the Canopus tracker once more locked onto a bright particle, sending the craft into an uncontrolled roll. Attempts to end the rolling resulted in the expenditure of yet more orientation gas. When only ten ounces of nitrogen remained from the original eight pounds—less than the amount engineers believed necessary for a successful third encounter—flight controllers decided to stop trying to stabilize *Mariner 10* and merely hold its motion within limits by periodically adjusting the solar panels. As the spacecraft drifted around the Sun on its way back to Mercury, mission controllers kept tabs on its orientation by noting the relative position of stars as they crossed the star sensors' field of view and by monitoring the waxing and waning strength of radio transmissions from the low-gain antenna. Three times in the next five months, they called upon the tiny reserves of gas to stabilize *Mariner 10* long enough for trajectory corrections. On March 7, the eighth and final burn of the mission put the probe on a course that would bring it within 203 miles of the dark side of Mercury.

On the last leg of its celestial marathon, *Mariner 10* ran true to form. Two days before the encounter, as the craft tried to lock onto Canopus again, the communications link with Earth was lost. In a desperate attempt to save the mission, ground control declared a spacecraft emergency and put in an urgent request for the services of the powerful Deep Space antenna in Spain, which was currently assigned to other duties. The signal from Spain went out and, with only hours to spare, *Mariner 10* located its guide star and was back on track. But then the cranky communications system threw its final tantrum. This time the faulty equipment was on Earth: The Australian Deep Space antenna that was to receive *Mariner 10*'s data suffered a cooling unit leak. With the antenna's receiving power reduced, controllers had to tell *Mariner 10* to transmit at the slower twenty-two-kilobit rate. Rather than compromise resolution by transmitting full images, team scientists programmed the spacecraft to return only a quarter of each. Luckily, the probe's whisker-close approach to Mercury ensured that its final pictures were some of its clearest.

The final rendezvous came on March 16. As *Mariner 10* swept toward the planet, its instruments again sensed the bow shock at the confluence of the solar wind and the magnetic field. The two magnetometers clearly registered the rising magnetism, indicating a field about a hundredth as strong as Earth's. Furthermore, the data indicated that the magnetism was intrinsic to the planet, not the result of exotic electrical effects. However, there was still no hint of how slow-turning Mercury might fit into the conventional model for planetary magnetic fields. Some experts felt that Mercury's field must emanate from permanently magnetized rock in its crust. Subsequent analysis convinced others that the field is created by Mercury's molten outer iron core, acting like a huge electrical generator.

THE RECORD AT JOURNEY'S END

By the end of the third encounter, *Mariner 10* had compiled an impressive record. Its cameras had scanned roughly half the planet's surface, yielding 2,700 useful images, some of which revealed features as small as 300 feet in diameter. Its instruments had relayed dozens of hours of readings, giving planetary scientists data they would use to develop scenarios of Mercury's geologic history.

One widely accepted theory of Mercury's evolution held that soon after the planet accreted from gas and debris in the primordial Solar System, an iron core coalesced under a fluid, melted mantle. At the same time, Mercury underwent an intense bombardment, possibly by bodies left over from the accretion. Heat generated by radioactive decay and the gravitational in-falling of matter to form the core caused the planet to expand, fissuring its crusty outer layer. Hot magma welled up through cracks, spilling out across great stretches of the planet. Eventually, Mercury began to cool and its crust thickened, causing planetwide contraction. Eruptions subsided as volcanic vents were closed off by the compressed crust, which threw up snaking lines of cliffs as it buckled. By the time *Mariner 10* arrived billions of years later, meteoritic pounding had turned Mercury's surface into a dusty, pitted desert.

The success of the Mercury mission could be measured by the fact that more than a decade later, planetary scientists were still combing through the sheaves of data from the probe, deepening and refining their knowledge of the planet. Space engineers also built on the legacy of *Mariner 10,* which included not only the gravity-assist technique but also the pioneering use of solar sailing for fuel conservation. The troublesome spacecraft that had done its job almost in spite of itself outlasted its mission by only a few days. On March 24, a short signal went out from JPL and crossed 130 million miles to *Mariner 10,* instructing the explorer to turn off its scientific instruments. The attitude-control gas had already run out, and for a while flight controllers tried to maintain stability with solar sailing alone. A short time later, however, the spacecraft lost its fixes on the Sun and Canopus and began a slow tumble. Saddened controllers sent a final command that turned off the transmitter. A billion miles and 506 days after launch, *Mariner 10* became a ghost ship, silently and eternally circling the Sun.

A CRATERED ARCHIVE

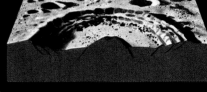

From above, large craters that are more than fifty-five miles in diameter often resemble shooting targets, with a circle of peaks rising from flat floors ringed, in the largest craters, by mountains.

Mid-size craters from six to fifty-five miles in diameter have flat floors, a terraced wall, and one or more central peaks.

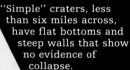

"Simple" craters, less than six miles across, have flat bottoms and steep walls that show no evidence of collapse.

When *Mariner 10* swooped past Mercury in 1974, the images it sent back revealed a planet that seemed at first glance to be a dead ringer for Earth's moon: Mercury's surface was pocked with craters and more craters, interspersed with smooth plains similar to the lunar maria and rugged areas reminiscent of the lunar highlands. But a closer look at the pictures revealed some striking nonlunar characteristics. The planet is laced with huge cliffs, or scarps, and vast regions known as intercrater plains. One extensive area is so hilly and lineated that scientists dubbed it the "weird terrain." The probe also detected a magnetic field, lending support to the theory that Mercury has an iron core.

To planetologists, these features suggest a geologic past distinctly different from the Moon's. This suspicion is bolstered by Mercury's cratering record, which leads many scientists to believe that the planet's surface is on average considerably older than that of any of the other terrestrial worlds, dating from the Solar System's infancy. Without an atmosphere or water to erode it, the surface has retained vital clues to the events that shaped Mercury, and perhaps the other terrestrial planets. Working almost solely from *Mariner 10* images, and without benefit of rock samples or data from manned or robotic spacecraft landers, scientists have applied remarkable sleuthing skills to reconstructing the history of what until fairly recently was the least-known member of the inner solar family.

As shown in these paired images, the dating rule of thumb is: Whatever lies beneath is older than what lies on top. At right, a class 3 crater named Rameau *(blue, bottom)* is deemed older than a scarp *(orange)* that transects it but younger than the cratered plain on which it is superimposed. The plains that are filling the craters postdate them but are older than the transecting scarp.

In the middle images, the youthful class 1 Kuiper Crater *(green)* shows a crisp rim, a pronounced ejecta blanket *(light green)*, and bright rays *(fine dots)*. Kuiper overlies the older, class 2 Murasaki Crater and its ejecta *(purple and light purple)*, which in turn top even more ancient plains.

At bottom right, a crater and its ejecta *(green and light green)* overlie three class 3 craters *(blue)* and a class 4 crater *(mauve)*. The crater is judged to be a young class 1.

PLAYING THE AGE GAME

Pinning down the exact age of a crater or lava field calls for the dating of actual rock samples, but photographs can help scientists determine their ages relative to one another. Clearly, for example, a lava field must be newer than any underlying craters it has partially destroyed but older than one whose outline in turn indents the lava. A widely used classifying system dates craters on a scale that ranges from class 1, the youngest looking, to class 5, the oldest looking. Young craters have distinct rims, prominent central peaks, bright rays, and well-defined ejecta splash marks. The features of old craters tend to be broken, covered, and worn away by such forces as lava flooding, later impacts, and the shrinking of the young planet's surface as it cooled. Judgments are not easy to make, however. For example, small craters tend to degrade faster from chipping and lava flooding than large ones and could be deemed older than they are.

Similar confusion can arise because of so-called secondary craters. Generally, the more cratered an area, the older it is judged. Its surface is assumed not to have been covered by recent geologic activity and thus would show the scars of eons of impacts. But when an object blasts out a primary crater, clods of ejected material pepper the surface and produce smaller, secondary craters that can falsely age the region as a whole. Only craters of similar size and degradation may be safely assigned similar ages. By counting craters of a given diameter in a region, scientists gauge the kinds of objects that made them and approximately when. Studies of the most heavily cratered areas on the Moon, for example, indicate that early in its life the Moon endured a heavy barrage from asteroid-size objects. The dating of Moon rocks has established that this era of heavy bombardment began about 4.5 billion years ago and lasted for some 700 million years. A similar period seems to have taken place on both Mercury and Mars, but astronomers are not certain that the three were simultaneous.

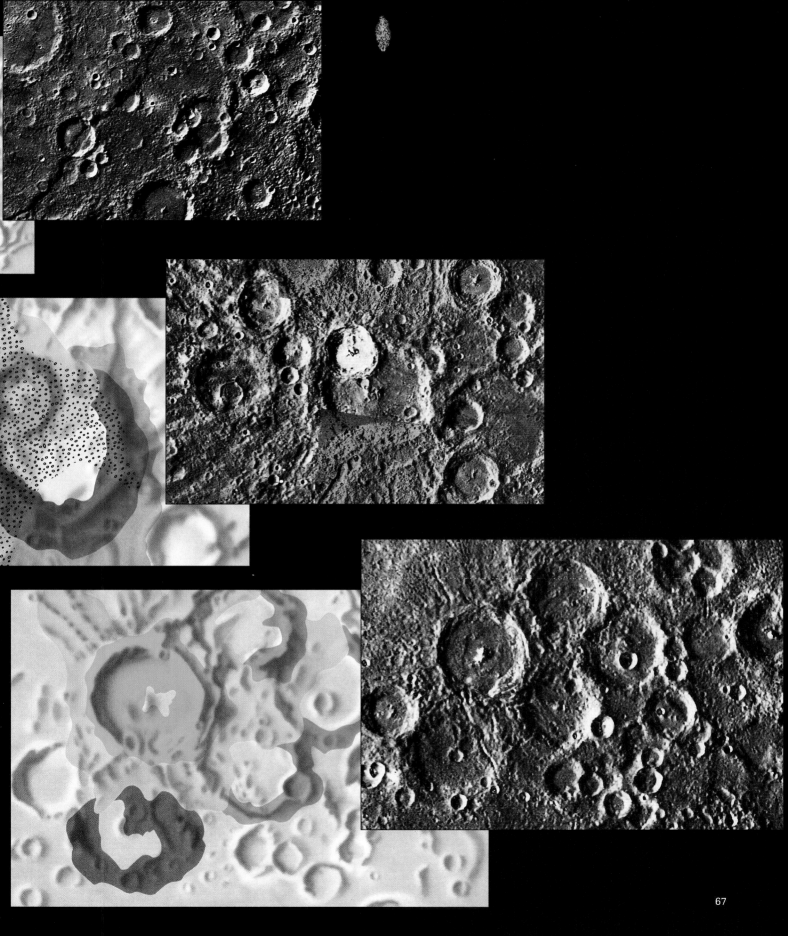

Flat to gently rolling intercrater plains *(below)* extend between and around clusters of large craters. The plains are heavily pockmarked with small, superimposed craters, most less than ten miles in diameter.

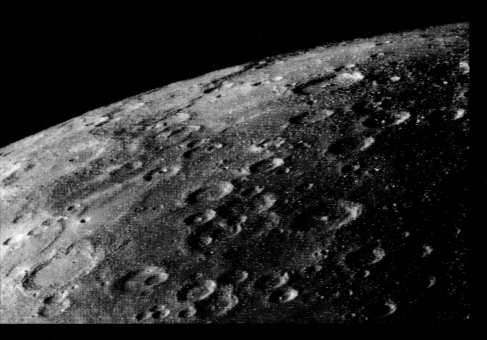

THE TIMEWORN PLAINS

Nearly half of the area of Mercury photographed by *Mariner 10* consists of terrain that scientists at first thought represented Mercury's original crust. Level or gently rolling, the intercrater plains, as these lands are known, thread between clusters of large craters and are pitted by innumerable smaller craters that bespeak long exposure and old age. Further study revealed that some plains overlie heavily cratered terrain and others have filled in huge craters called basins. The intercrater plains still rank among the most ancient of Mercury's terrains, but their features are now believed to have been formed at different times during the period of heavy bombardment, and perhaps by different means.

Any or all of three processes could have contributed to their formation. Giant impacts that blasted out basins early in Mercury's history may have thrown up quantities of debris that rained down, slowly filling in old craters and leveling the landscape. But scientists cannot find enough old basins to account for the material needed to form all of the plains. Another scenar-

io suggests that after Mercury's accretion into a planet, repeated heavy bombardment plowed the surface. Melted by global heating, the surface settled and eventually solidified into level plains, which were later cratered in a final episode of intensive bombardment. The third theory, favored by many, is that lava flows covered large areas, obliterating smaller craters. This could be why, compared with similar areas on the Moon, Mercury's heavily cratered highlands are marked by few craters less than thirty miles in diameter. But *Mariner 10*'s cameras could not resolve smaller features well enough to allow scientists to positively identify landforms, such as cones and lava flow fronts, that would indicate volcanism in Mercury's past.

In an artist's conception *(right)*, meteoroids vaporize on impact, ejecting material that may now be part of the intercrater plains.

In the photograph at far right, Vostok scarp cuts through the crater Guido d'Arezzo. The cutaway view at right reveals the cliff to be the exposed, beveled edge of a section of Mercury's crust that has been thrust up along a fault line and over the adjoining section. The elevated crust includes the northeastern portion of the crater and its rim, which has been offset by six miles.

This trio of drawings depicts a widely accepted theory of events leading to scarp formation on Mercury. Shortly after the planet formed by accretion (above), temperatures began rising from heat generated by the accretion process, by radioactive decay, and by tidal effects that tended to brake the planet's spin. Gravity pulled heavy iron molecules inward, separating them from rocky silicates, which contributed to the temperature increase.

By the time Mercury's large iron core (orange), mantle (yellow), and crust (brown) had formed, planetary temperatures had risen high enough to cause the mantle to begin to melt. This heating also had the effect of expanding the globe's radius by three to six miles, increasing the surface area by perhaps two percent and fracturing the thin crust.

Looming Ramparts

Travelers crossing Mercury's bleak landscape would often find their way barred by awesome cliffs, or scarps, rising from a few hundred yards to two miles into the sky and extending over distances ranging from 12 miles to more than 300. Known as lobate scarps, because they follow paths that trace the outlines of giant lobes, the cliffs cut through craters and snake across smooth plains and intercrater plains—features that must therefore all be older. Yet the scarps themselves are dotted with class 1 and class 2 craters, obviously punched out after the cliffs had formed. Since the transition from class 3 to class 2 craters is believed to mark the end of the era of heavy meteorite bombardment, the scarps may have begun to form then.

Mercury's high density long ago led to a hypothesis that the planet had a massive iron core, a theory supported by *Mariner 10*'s discovery of the planet's magnetic field. Formation of the core, along with planetary accretion and other factors, would have generated such vast amounts of heat *(diagrams, left)* that the planet expanded, causing the thin crust to crack. Eons later, as global cooling set in, the planet began to contract. The scarps' configuration suggests that this cooling compressed Mercury's crust, thrusting great slabs over adjacent areas to build ridges—a process that may still be going on.

As Mercury cooled, the crust thickened and the mantle and part of the core solidified, causing the planet's radius to shrink by about a mile and decreasing its surface area by more than one and a half percent. Buckling sections of the crust broke along fault lines and were thrust over adjoining regions. The protruding edges of the upthrust sections produce lobate scarps.

As shown in the globe diagram at left, the meteorite impact that formed the Caloris Basin generated surface waves that traveled around Mercury through the crust, while compression waves raced through the dense core. Meeting and reinforcing at a point directly opposite the impact, they created the weird terrain, a region nearly as large as West Germany and France combined. The photograph above shows the area's irregular hills, some as much as a mile high, and a webwork of valleys, some more than seventy-five miles long and nine miles wide.

ANCIENT DEVASTATION

Some four billion years ago, Mercury was struck a cataclysmic blow. A giant meteoroid, perhaps as much as 100 miles wide, slammed into the young planet, blasting a gargantuan hole in the crust and sending up untold tons of molten and fragmented material. The shards rose miles above the surface, then rained down near and far, the larger fragments causing further devastation. Mute testimony to that stupendous event is the Caloris Basin, a crater more than 800 miles in diameter with a rim more than a mile high. Were a similar meteoroid to hit central France, the crater and a surrounding blanket of debris, as much as one and a quarter miles thick, would cover most of Europe. City-size secondary craters would pockmark outlying areas of the continent, with smaller excavations occurring as far away as the Soviet Union, Turkey, and northern Africa.

On Mercury, the consequences of the impact did not stop at the Caloris Basin. Seismic waves generated by the blow went rippling around and through the planet to meet and reinforce each other on the opposite side of the globe from the impact site. Within minutes, the resulting shock raised the surrounding surface at least a mile, fracturing the crust tens of miles deep and tearing up and jumbling the landscape into the choppy region named the weird terrain. More than 223,000 square miles, the area is, like the Caloris Basin, speckled primarily with class 1 and class 2 craters, lending credence to the theory that the two features are related and were formed together near the end of the heavy bombardment.

Smashing into Mercury with the energy equivalent to a trillion one-megaton hydrogen bombs (right), an immense meteorite blasts out gigantic Caloris Basin, sending powerful shock waves through the planet and lofting enormous quantities of debris.

Signs of Youth

Vast sections of Mercury resemble the lunar maria, the darker, sealike areas that form the "man in the moon" image. Mercury's counterparts—relatively flat, low-lying, and sparsely cratered—are called the smooth plains. Most scientists think they are flood basalts of volcanic origin, like the lunar maria, although *Mariner 10* images show few definite volcanic landforms. The basaltic lava might have welled up from fissures that either remained in the crust after the planet's expansion or were reopened by the impacts of meteorites.

Scientists believe that the smooth plains constitute the planet's youngest surface: Not only do they fill large craters (which must have formed first), but they themselves have not been marred by many large craters. Yet the craters that do appear on the smooth plains are similar in size and number to those in the heavily cratered lands created on the other terrestrial planets during the period of heavy bombardment. Scientists thus conclude that even though the smooth plains are the youngest type of terrain on Mercury, they were formed some 3.8 billion years ago. Hence, the planet's surface, relatively undisturbed since, is on average older than that of any other terrestrial world.

As rendered by an artist, lava gushing from a large fissure in Mercury's crust *(left)* flooded ancient basins and buried small crevasses. The molten basalt eventually solidified into a smooth plain, similar to the surface that is southeast of the Caloris Basin shown below. The plains are also marked by a wrinkle ridge created by the subsidence of the crust under heavy lava deposits.

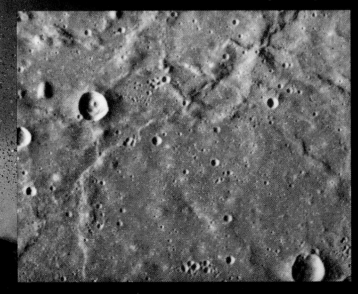

The golden crescent of a Venusian dawn burnishes the dense veil of carbon dioxide clouds that hides a world of bleak highland regions, vast depressions, and volcanic formations unlike any on Earth.

iamond-bright in the morning and evening skies, Venus is one of the most exquisite of celestial objects. Its light is brilliant enough to cast a shadow at night and to render the planet visible even by day. Three factors make Venus so luminous. Two of them—the planet's proximity to Earth (about 24 million miles away, at the closest) and its relatively large size (almost that of Earth)—set it up as a natural target of astronomical investigation. The third makes it an astronomer's nightmare: Venus is blanketed with highly reflective clouds, masses of yellow-white vapor that reflect 76 percent of the sunlight falling on them. As a result, the planet's surface is invisible to optical instruments. The clouds stymied scientists until the twentieth century, when astronomers began to observe the planet in a wider range of the electromagnetic spectrum. Even so, progress was slow, largely because the slowly rotating planet shows essentially the same face to Earth whenever it is near enough for good observation. But in 1957, the then-Soviet Union launched a satellite named Sputnik, inaugurating the space age. Unmanned missions, including the recent *Magellan* orbiter's observations, have since allowed scientists to create a reasonably accurate profile of Earth's nearest neighbor.

Like the Mercury missions, the Venus programs were developed against the backdrop of the emerging space race. By 1962, both the Soviets, who had launched the first person into space, and the Americans, who had committed themselves to a Moon flight, were looking for ways to demonstrate their growing prowess in the new field. Venus was an early goal.

Initially, the two countries took different approaches. Soviet planners favored an aggressive and ambitious program that aimed to penetrate Venus's atmosphere and land on the surface. For this reason, their vehicles were insulated, bulky, steel-plated behemoths weighing anywhere from 1,400 to more than 11,000 pounds. Americans started with craft that would stay safely above the cloud layer and examine the atmosphere and surface with remote sensors. These were neat, delicate machines weighing about 450 pounds.

Despite their divergent technologies, the two nations had similar questions in mind as they began to explore Venus. Both were interested in the nature and geology of the Venusian surface. Both wanted to know if the planet had a magnetic field. Perhaps the most fascinating subject of all, though, was the Venusian atmosphere. Scientists hoped to learn not only what chemicals make up the dense clouds but also whether a key substance—water vapor—might be found there, possibly testifying to ancient oceans like those on Earth.

INVASION FROM EARTH

The first interplanetary probes were disappointments, however. The Soviets' *Venera 1* (Venera is the Russian word for "Venus"), launched on Februrary 12, 1962, lost contact with ground controllers when it was less than five million miles from Earth—only a fifth of the way to Venus. A radio link was never reestablished and its fate remains unknown. On July 22, 1962, the U.S. launched *Mariner 1*, replete with hardware designed for the earlier, Moon-surveying Ranger program. But disaster struck almost immediately. In the skies over Cape Canaveral, the Atlas rocket carrying the vehicle went off course. Less than five minutes into the flight, controllers intentionally destroyed the craft. Back then, though, U.S. space authorities built in pairs, and thirty-six days later, *Mariner 2* soared into space. On December 14, 1962, the vehicle gave a major boost to the American program as it flew within 22,000 miles of Venus's hidden surface. In doing so, *Mariner 2* became the first probe to successfully approach any planet beyond Earth.

The expansion of astronomers' small fund of knowledge about Venus was sudden and dramatic. *Mariner 2*'s infrared detectors disclosed a layered cloud structure. To the surprise of the mission's designers, magnetometers aboard the probe detected no sign of a magnetic field around the planet. Venus's density and size had led scientists to presume that the planet possessed a core structure similar to Earth's—that is, a solid inner core surrounded by a fluid outer one. If so, electrical currents generated in the outer core by planetary rotation should have produced a significant magnetic envelope. The absence of one meant astronomers had to revise their theories of Venusian anatomy. *Mariner 2* also measured atmospheric temperatures near the surface of 750 degrees Fahrenheit—far too high for the existence of liquid water even under very high atmospheric pressure.

Just over a year later, the Soviets returned to Venus when *Venera 2* flew within 15,000 miles of its target on February 27, 1966. Inexplicably, the craft did not return any data. *Venera 3* was on course to land on Venus when, like *Venera 1*, it lost its communication link to Earth. Controllers concluded that it crashed to the surface on March 1.

The next launching opportunity arrived in June of 1967. By that time, space engineers in both countries were working feverishly on Venus missions. The Soviets got off the ground first. On June 12, they launched *Venera 4*—at 2,400 pounds the biggest interplanetary spacecraft yet. Two days later, the U.S. launched *Mariner 5,* a spare from the 1964 Mariner program that had explored Mars. The two craft reached Venus almost simultaneously in October 1967.

Upon arrival, *Venera 4* transmitted data for one hour and thirty-four minutes before it faded out. Later, faint but historic signals reached Earth from a descent probe ejected by the main vehicle. For the first time, a spacecraft had entered another planet's atmosphere and lived long enough to tell the tale.

The probe indicated that temperatures in the Venusian clouds ranged from 100 degrees to 525 degrees, that the planet's surface pressure exceeded that of Earth by at least a factor of fifteen, and that the atmosphere was made up of 98.5 percent carbon dioxide. The probe did not detect any evidence of a magnetic field or radiation belts as it fell, presumably to a fatal crash landing.

Meanwhile, *Mariner 5* flew past the planet at a distance of 2,100 miles. The angular little craft, smaller than a compact car, found no hint of a magneto-sphere, but it did find a bow shock wave of charged particles on one side of Venus, caused as the planet's electrically conducting atmosphere deflected the solar wind around the planet. Controllers were able to combine *Mariner 5*'s tracking data with Earth-based radar measurements to determine that Venus is more nearly spherical than Earth. Even its flight path was inform-ative: Because the gravity of Venus had affected the craft's trajectory by a measurable amount, *Mariner 5* yielded a more precise estimate of the planet's mass than had been possible before. (The new figure was 0.8149988 times Earth's mass.) But frustrated scientists wanted to know much more. How deep were the clouds, and what were they made of? How did the atmosphere circulate around the planet? Did the atmosphere contain water and oxygen? What did the planet's surface look like?

An International Cast of Explorers

Representing more than a quarter-century of Venus exploration, the American and Soviet probes shown at right and on the next two pages have run up an im-pressive list of achievements. The five U.S. missions began in 1962 with *Mariner 2*, which made the first up-close measurements of Venus's temperatures and magnetic field. *Mariner 10* returned the first photos of Venus taken from space. The 1978 Pioneer Venus Or-biter and Pioneer Venus Multiprobe missions sent back high-resolution surveys of the planet's terrain, deploying four probes into its thick mantle of clouds.

The Soviet effort, with eighteen missions, is a litany of firsts. *Venera*s 7 and 8 in the early 1970s transmit-ted more than an hour of data from the furnacelike surface, and in 1975 *Venera*s 9 and 10 became the first probes to photograph another planet from the ground. *Venera*s 15 and 16 radar mapped Venus with ten times Pioneer's resolution in 1983. *Vega*s 1 and 2 deployed French-developed balloon-borne packages, the first sensors to float in an extraterrestrial atmosphere.

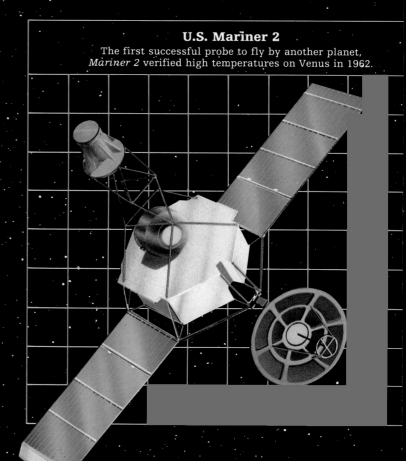

U.S. Mariner 2
The first successful probe to fly by another planet, *Mariner 2* verified high temperatures on Venus in 1962.

Delving into these issues required craft that could tolerate the Venusian environment for a longer time. The Soviets met the challenge with their beefed-up *Venera 7* spacecraft, which reached Venus in December 1970. Its 1,100-pound landing capsule had extra insulation so that it might withstand, at least temporarily, the fall through the planet's brutal temperatures and atmospheric pressures. The lander was also equipped with heavy shock absorbers for its encounter with the ground. Detached from the mother ship, the capsule dropped toward the surface by parachute. Because no probe had ever survived a descent before, ground controllers were not even sure how long the trip would take. But after thirty-five minutes, the scientists noticed that their readings of the probe's speed and of the temperature around it suddenly leveled off; they continued to receive data for twenty-three minutes more, until contact with the probe was lost. The jubilant conclusion: The probe had indeed landed on the surface of Venus. There, it recorded a temperature of 880 degrees and a pressure ninety times that on Earth's surface.

Within two years, the Soviets topped their own performance. On July 22, 1972, *Venera 8* touched down on the day side of the planet and radioed back a stream of data for an unprecedented fifty minutes. Among other things, the vehicle detected traces of ammonia in the atmosphere and revealed a chang-

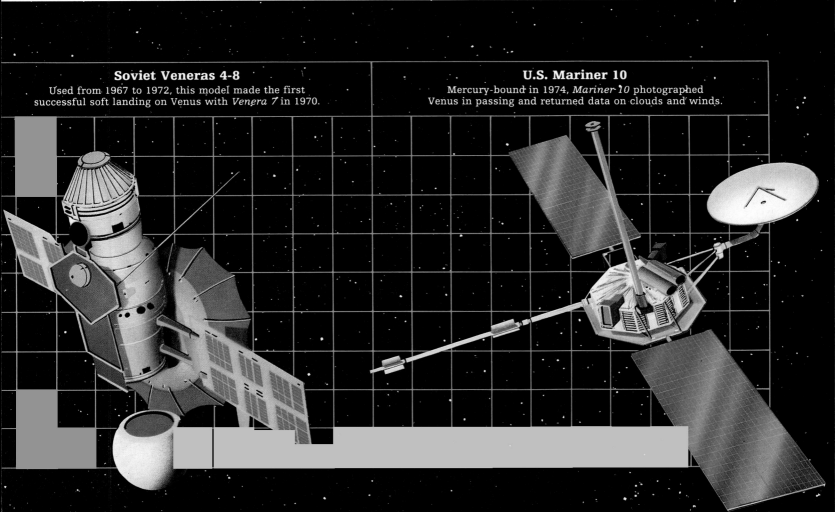

Soviet Veneras 4-8
Used from 1967 to 1972, this model made the first successful soft landing on Venus with *Venera 7* in 1970.

U.S. Mariner 10
Mercury-bound in 1974, *Mariner 10* photographed Venus in passing and returned data on clouds and winds.

ing pattern of Venusian winds, with gusts of 300 feet per second at high altitudes, calming to three feet per second close to the surface. *Venera 8*'s instruments also measured the density of the surface materials at 1.5 grams per cubic centimeter, similar to that of dirt on Earth. Like terrestrial soil, the Venusian dirt was slightly radioactive, containing significant amounts of uranium, as well as thorium and potassium. Perhaps most important was the finding that, despite its heavy cloud cover, the Venusian surface was at least dimly illuminated. Of the solar energy that reached the cloud tops, about two percent was getting down to the surface—similar to the amount that reaches Earth's surface on a very overcast winter's day. Engineers were delighted. If a lander could survive on the surface of Venus, then it should be able to produce video pictures of its surroundings.

While Soviet engineers worked on building a suitably sturdy camera, Americans prepared to snatch another cloud-top view of Venus from Mercury-bound *Mariner 10*. En route to its real target, the craft would first swing around Venus to gain a boost from Venusian gravity. On February 5, 1974, *Mariner 10* began to broadcast the first images of Venus ever produced from space. At the Jet Propulsion Laboratory, which housed mission control for the Mariner Mercury project, scientists watched monitor screens in wonder as the

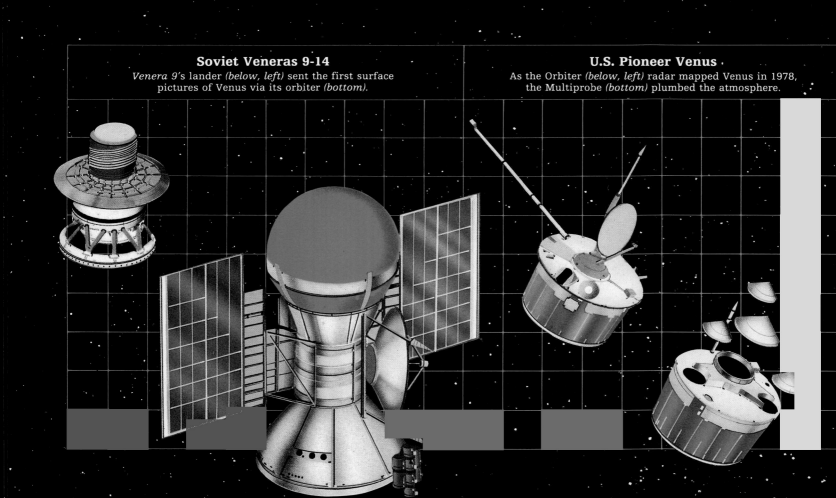

Soviet Veneras 9-14
Venera 9's lander *(below, left)* sent the first surface pictures of Venus via its orbiter *(bottom)*.

U.S. Pioneer Venus
As the Orbiter *(below, left)* radar mapped Venus in 1978, the Multiprobe *(bottom)* plumbed the atmosphere.

pictures, shot in ultraviolet light to pick out patterns most easily, portrayed the swirling clouds. The astronomers recognized the patterns as closeups of Y- and C-shaped markings seen fuzzily by Earth-based telescopes.

The detail in the images provided new information about Venus's roiling shroud and allowed observers to fine-tune their ground-based observations. Still, scientists coveted images from the hidden surface. Despite the high temperatures and hostile atmosphere—discouraging evidence of a barren world—some observers still hoped for a little swampiness, a puddle amid heated rocks. Toting digitized camera equipment, two Soviet Venera craft made their way toward Venus.

On October 22, 1975, *Venera 9* touched down on the surface of the northern hemisphere and, for fifty-three minutes, transmitted data that produced the first portrait of the Venusian landscape. The historic black-and-white image showed a dark, forbidding plain, strewn with sharp-edged, flat rocks, extending to a nearby horizon. The picture contained no hint of water, present or past. Geologists identified the stones as basalts—igneous rocks that had erupted from the planet's interior. On Earth, basalts cover the ocean floors and parts of continents. The evidence seemed to confirm radar studies identifying the area as volcanic.

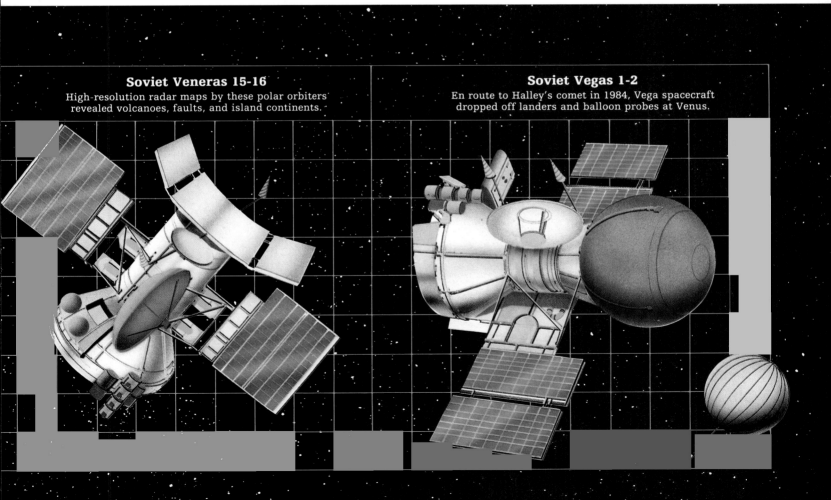

Soviet Veneras 15-16
High-resolution radar maps by these polar orbiters revealed volcanoes, faults, and island continents.

Soviet Vegas 1-2
En route to Halley's comet in 1984, Vega spacecraft dropped off landers and balloon probes at Venus.

Three days later, *Venera 10* landed about 1,400 miles south of *Venera 9* and sent back a second view of the surface. The image showed a different, more rounded type of rock and revealed greater areas of soil-like debris between individual rocks, apparently the result of greater chemical and wind erosion. Again, there was no sign of water.

THE GREENHOUSE EFFECT

This extreme environment, more desiccated than Earth's driest desert and hot enough to melt zinc, begged for an explanation. With its cloak of clouds, the planet should actually reflect most of the Sun's light, leaving its surface relatively cool and more like that of Earth. How could such hellish conditions have developed? Carl Sagan was sure he knew the answer.

In the early 1960s, when he had a joint appointment at the Smithsonian Astrophysical Observatory and Harvard University, Sagan started fundamental research into the surface and atmosphere of Venus. Dismissing many rival notions about soda-water oceans and the like, as well as arguments that temperature measurements from Venus had to be wrong, he revived an old theory about the so-called greenhouse effect, proposed by German-American scientist Rupert Wildt in the 1940s. Wildt had suggested that the clouds of Venus trap heat much as do the glass walls of a greenhouse.

According to Wildt and later Sagan, Venus may have started out as a colder planet with a dense, watery atmosphere composed of gases vented by the cooling crust. But because energy from the Sun was partially trapped by the clouds, the surface temperature rose inexorably, eventually baking carbon dioxide vapor out of the very rocks. The increasingly dense carbon dioxide atmosphere absorbed and reradiated long-wavelength infrared energy from the planet's surface, continuing to raise surface temperatures until they reached the level seen today.

Sagan believed that water vapor was essential to this equation. With only a carbon dioxide atmosphere, enough surface radiation could leak back up through the gas to keep the temperature from rising to the level measured by space probes. A layer of water-ice clouds atop an atmosphere containing water vapor, he calculated, would absorb sufficient infrared radiation to make the Venus greenhouse work.

Sagan's colleagues were dubious. No evidence existed to support his idea of icy clouds. But Sagan was undaunted, and by 1973, the tide had begun to turn his way. That year, at a meeting of a division of the American Astronomical Society in Tucson, Arizona, several scientists suggested that the clouds of Venus might consist of highly corrosive sulfuric acid. This hypothesis, based on data from ground-based studies, gave Sagan a boost, even though it contradicted his water-ice idea. Like most naturally occurring acids, the sulfuric compound would most likely be held in a water solution. And that water, in addition to carbon dioxide, should be enough to create an extreme, or "runaway," greenhouse effect on Venus.

Conveniently for the scientists, both the U.S. and Soviet space programs were planning their biggest assault yet on Earth's sister world. Scheduled for a Venusian rendezvous in 1978 were two American craft and two Veneras. One of the U.S. probes would orbit safely above the planet's atmosphere. The rest would take the suicidal path to the surface, radioing back critical facts as they plunged to their doom.

American scientists and engineers had planned the double-barreled mission known as Pioneer-Venus for more than seven years. Focusing its investigation on the planet's gaseous shell, the Pioneer team decided to look closely at the composition, layering, and circulation of the clouds; monitor solar heating of the atmosphere; record changes of temperature with changing altitude; and measure wind speeds at various heights. In addition, the group decided to seek as complete a view of the planet's topography as possible.

The first vehicle, the Orbiter, had as its most important cargo a twenty-four-pound radar mapper, designed to transmit topographical information about the surface of Venus as the craft swung around the planet once every twenty-four hours. Radar maps are very different from photographs, and require a very different kind of interpretation. Unlike a camera, in which film simply registers the visible light emitted or reflected by a surface, a radar device aboard a spacecraft bounces radio waves off a planet's surface and measures the time it takes for the waves to return; differences in return times reveal the elevation of various features on the surface. The device also measures the intensity of the return signal, which yields information on surface roughness and average slopes; generally, the rougher the surface and the steeper the angle of incidence, the brighter the radar image.

A QUARTET OF PROBES
The second Pioneer was the Multiprobe, consisting of a mother ship, or bus, carrying four instrument-packed probes. First the probes, then the bus, would drop into the murk. To withstand the extreme heat and pressure as long as possible, the probes were made of solid titanium and tightly sealed; their windows—necessary for admitting light and heat for experiments—were milled from diamond and sapphire. The Multiprobe's grab bag of sophisticated, miniaturized instruments featured a mass spectrometer and gas chromatographs for measuring the composition of the atmosphere, and a net flux radiometer to find out how and to what extent clouds absorbed solar radiation. None of the probes were specifically designed to survive the impact with Venus, although engineers remained hopeful that their wide, cone-shaped heat shields would slow their descent through the dense atmosphere and allow them some chance of cheating death, if only for a few moments.

When the time came, the Orbiter entered an elliptical orbit around Venus to begin its mapping chores, and the four sacrificial probes made their perilous dive toward the rocky world below. As they fell, transmitting information about the smog around them, a crucial device—the mass spectrometer aboard the largest of the four probes—cut off suddenly. The ground crew frantically tried to diagnose the problem. Before they found the answer, the signal began to return, gradually building up to full strength. Relieved but mystified, the scientists went back to tracking all four probes' transmissions. Nearing the surface, the probes, like their Soviet predecessors, registered Venus's blistering heat and the rising atmospheric pressure. Then, one after the other, they smashed into the ground at about twenty miles an hour. Three stopped transmitting; the fourth somehow survived and continued to send signals for sixty-eight minutes after impact. Finally, its batteries died.

PIERCING THE VEIL

Analysis of the Multiprobe signals divulged a wealth of information about the planet. At high altitudes, Venus's atmosphere turned out to be cooler than Earth's, perhaps because so little surface heat escapes to those levels. Beneath the clouds, the temperature at a given height was relatively constant across the globe. In fact, the region below the clouds was surprisingly stable, with none of the turbulence suggested by the winds raging aloft. The atmosphere seemed to consist mainly of carbon dioxide and nitrogen, with little free oxygen or water vapor. The mass spectrometers also measured unexpectedly high abundances of the inert gases neon and argon 36 and 38, suggesting that Venus and Earth received different amounts of volatile compounds from the Solar nebula out of which all planets are thought to have condensed.

Two weeks after the Pioneer spacecraft arrived at Venus, *Venera*s *11* and *12* each dropped an entry probe as they flew by. Among other things, the Soviet probes detected what may have been intense thunderstorms producing twenty-five lightning impulses a second. They also confirmed the presence of a small amount of water vapor in the lower atmosphere of the planet.

For American scientists, the Venera data was eclipsed by a belated dis-

Under a sky that filters blue light, *Venera 14*'s spiky foot *(bottom)*, a discarded lens cover *(center)*, and a color gauge *(right)* provide scale and hue references for an orange landscape. In 1982, the Soviet probe and its twin, *Venera 13*, landed about 600 miles apart south of Beta Regio, the smallest of three known Venusian landmasses. Soviet scientists believe this region was once covered with a thin layer of lava that fractured upon cooling to create a surface marked by rounded, interlocking shapes. Soil samples analyzed by Venera's robot laboratory revealed a composition similar to that of basaltic lava rocks on the terrestrial ocean floor.

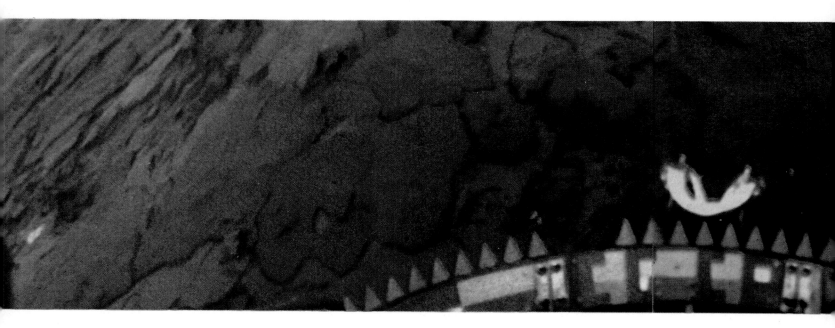

covery resulting from the temporary failure of the mass spectrometer on the largest of the four Pioneer probes. Upon analysis, scientists learned that an inlet on the instrument had been clogged by a small drop of liquid, despite the fact that engineers had installed a heater coil around the area to boil off any such intrusion. In fact, the heater had worked, but not as fast as expected. When the droplet did evaporate, what it left behind caused considerable excitement. There was evidence of both sulfur—most likely the residue of sulfuric acid—and deuterium. Measurements have since shown that acid in the clouds is more concentrated than that in a typical car battery. The presence of water in the sulfuric acid solution, the discovery of water vapor by *Veneras 11* and *12,* and an accumulation of observations since then have apparently vindicated Carl Sagan's theory of a carbon-dioxide/water shield in Venus's atmosphere. While carbon dioxide is responsible for the bulk of the greenhouse effect on Venus, water vapor is the next most important agent, followed by sulfuric acid and sulfur dioxide gas.

The deuterium discovery seemed in some ways even more significant, for it hinted at the possibility of large quantities of water on Venus, at least in the past. On Earth, deuterium, also known as heavy hydrogen, is found in tiny quantities in ocean water, along with light hydrogen and oxygen. The amount found in the inlet was disproportionately high, suggesting that the deuterium was all that remained of a Venusian ocean after the normal hydrogen and oxygen had vanished. Perhaps, said some theorists, Venus once had not just a steamy atmosphere but liquid water on its surface as well. The increasing heat of the greenhouse effect could have vaporized the planet's oceans; ultraviolet radiation from the Sun might then have split the water molecules into oxygen and heavy and light hydrogen. The lighter molecules of hydrogen would have readily escaped into space, leaving behind the heavier deuterium. If the greenhouse effect explained the disappearance of Venusian oceans, one could assume that Venus had resembled Earth when both were young—a scenario with huge implications for Earth's future.

However, not all astronomers accepted evidence of heavy hydrogen in the present as proof of oceans in the past. Some argued that we know too little about the formation of the inner planets to assume that the original ratios of normal and heavy hydrogen were the same on Earth and Venus. Lacking further signs of

erstwhile oceans, the question remained unresolved, as did Sagan's greenhouse model of the planet's atmospheric evolution.

Over the next few years, scientists looked for clues to Venus's past in the planet's topography, as revealed by the Pioneer-Venus Orbiter and a follow-up quartet of Soviet craft in the mid-1980s. Although Earth-based radar in the early 1970s had sketched out a few large-scale features, Pioneer's altimetry readings from Venus orbit unveiled a breathtaking landscape. In the north, a plateau larger than Tibet rises more than a mile above the surrounding plain, ringed by belts of mountains that include a range with a peak nearly seven miles high. The plateau, named Lakshmi Planum after the Hindu goddess of fortune, and the mountains together make up a continent-size highland region called Ishtar Terra after the Babylonian goddess of love and war. (The International Astronomical Union generally labels features on Venus with the names of classical goddesses and famous women. Three exceptions to the rule are Alpha Regio and Beta Regio, highland regions that were the first features to be discerned by Earth-based radar, and Maxwell Montes, the high peaks Pioneer discovered on Ishtar Terra, named for Scottish physicist James Clerk Maxwell.) Southeast of Ishtar and just south of the equator lies Aphrodite Terra, a highland nearly the size of Africa and marked by long depressions that resemble terrestrial rift valleys. South of Ishtar and west of Aphrodite, Pioneer detected two peaks on Beta Regio that looked volcanic.

In the spring of 1981, having mapped about 92 percent of the Venusian globe, the Pioneer Orbiter ceased radar transmissions. But a closer look at the surface came the following March, when *Venera*s *13* and *14* descended to soft landings and photographed their surroundings in both black-and-white and color. Because the thick atmosphere removes blue from the sunlight passing through it, the color images showed an orange sky and a fiery glow over the ground. Between large, thick boulders of dark gray rock lay a brownish black, fine-grained material. Rudimentary on-site analysis of soil samples revealed that the surface consisted of the same kind of basaltic rock seen on earlier missions, adding to growing evidence that Venus was, at one time at least, highly volcanic.

In October of 1983, the *Venera 15* and *16* orbiters took over Pioneer's imaging of the planet. The pair improved by a factor of as much as 30 the resolution of the portrait drawn by Pioneer, distinguishing surface features as small as seven-tenths of a mile across. Maps based on data from the Venera 15/16 mission depicted what seemed to be calderas, lava flows and other evidence of abundant volcanism, as well as ridges and troughs signifying complex tectonic deformation.

All in all, the spaceborne images were tantalizing—and frustrating—leaving planetary scientists with more questions than answers. For example, none of the probes had found the extensive impact cratering characteristic of Mercury, Mars and the Moon, implying that the Venusian surface is relatively young. But how recent was the activity that might have covered over the scars of ancient bombardments? And was that activity largely volcanic, or has the surface of Venus been shaped by the kind of processes— namely, plate tectonics—that have defined Earth's profile? And of course, there was the long-standing issue of Venusian oceans. Could a more detailed look at the surface pick out signs of ancient shorelines?

Any such examination lay years in the future. Except for the fleeting glances afforded by two flyby Soviet craft, *Vega 1* and *Vega 2*, en route to encounters with comet Halley in 1984, the surface of Venus once again became the sole province of Earth-based radar. Through the 1980s, that tool grew increasingly powerful, producing images that rivaled the closeup detail returned by Venera 15/16. Then, as the twentieth century entered its last decade, the United States sent another robot ambassador to Earth's sister.

The Magellan mission was the first in a series marking the resumption of American planetary exploration after a twelve-year hiatus dating from the launch of Pioneer-Venus. The project, originally known as Venus Orbiting Imaging Radar, or VOIR, had been on the drawing boards at the Jet Propulsion Laboratory since the early 1970s. Canceled in 1982, VOIR was scaled back and then reincarnated in late 1983 as the Venus Radar Mapper (VRM), on condition that the spacecraft be built for about half the original cost.

The new vehicle, later renamed for the sixteenth-century Portuguese explorer Ferdinand Magellan, made use of proven technologies and a host of hand-me-down parts. Its 3.7-meter high-gain antenna, used for both radar mapping and data communications with Earth, was a spare from the Voyager mission, as were the low-gain antennas, the main body of the craft, and its small thrusters. From the Galileo mission, scheduled to head for Jupiter in 1989, came a spare command and data system and attitude control computer. The medium-gain antenna, a backup for communication with Earth, was left over from the 1971 *Mariner 9* project to Mars.

Then, with all on track for a May 1988 launch and a short, four-month flight to Venus, *Magellan* was derailed again. The *Challenger* disaster in 1986 was followed by a thirty-two-month suspension of shuttle activities that resulted in increased sensitivity to safety considerations and much reshuffling of planned space missions. When the dust settled, *Magellan* had a new upper stage booster, the U.S. Air Force's Inertial Upper Stage (IUS), and a new launch date, May 1989. This launch window allowed the IUS, which was considerably less powerful than the Centaur rocket *Magellan* was to carry originally, to put the craft on a trajectory that would take it one and a half times around the Sun before arriving at Venus fifteen months later.

Released into space on May 4, 1989, by astronauts aboard the space shuttle *Atlantis,* the long-awaited robot explorer enjoyed a relatively uneventful

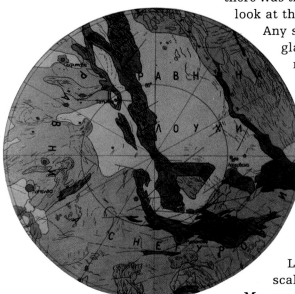

In the first image of Venus's north polar region *(above),* taken by *Veneras 15* and *16,* terrain disrupted by impact craters *(violet),* volcanic calderas *(orange),* and mountains *(blue)* suggests that the planet has undergone a long period of active crustal folding and faulting.

A radar map *(opposite)* assembled from *Venera 15* and *16* data shows the contours of a plateau named Lakshmi Planum on Ishtar Terra, Venus's northern continent. Ringed by mountain ranges *(brown)* and marked by two volcanic craters *(orange)* at its center, Lakshmi Planum covers nearly 640,000 square miles, twice the area of the Tibetan Plateau, with peaks that soar two miles high.

cruise, giving controllers back at JPL the chance to shake down its various systems and components and to evaluate—and, if necessary, revise—their plans for operations once the probe arrived at Venus. For example, engineers tested and calibrated *Magellan*'s star scanner, critical to keeping the craft pointed accurately during different phases of the mission, and the gyroscopes that feed information into the attitude-control system. With only minor exceptions, the cruise tests delivered no surprises.

On August 10, 1990, after a journey of 950 million miles, *Magellan* arrived within sixty miles of its aim point. As planned, the craft decelerated and inserted itself into a highly elliptical orbit around Venus. But then, as engineers back on Earth began the scheduled in-orbit check, the trusty probe went AWOL. On the 16th of August and again on the 21st, radio receivers at home lost the craft's signal. For some inexplicable reason, *Magellan* had turned its high-gain antenna away from Earth. Ground controllers both times managed to reestablish contact within a day, but despite extensive investigation, they could not determine why the craft had changed its pointing direction.

Even as the engineers tinkered with the craft's computer software to minimize damage if the problem occurred again, mission scientists were marveling at *Magellan*'s first radar images of its target. Orbiting as close as 182

Penetrating the Veil of a Goddess

Magellan's sophisticated radar imaging equipment has revealed the face of Venus in stunning detail. The scenes produced from the probe's data are not photographs, however, and are therefore subject to a different interpretation. Because very little sunlight gets through the Venusian atmosphere, brightness in an image is not a function of light but of radar reflectivity. As a general rule, the greater the angle of incidence of the terrain to the incoming radar, the more reflective it is. (Angle of incidence depends on the position of surface features in relation to the moving craft.) Elevation and the chemical composition of rock and soil also affect brightness.

As shown here and on the following pages, the Venusian surface is the product of a tortured geological past, one similar to Earth's in some respects and vastly different in others. Scientists are now bent on understanding the processes that helped create the planet's complex visage.

Released by the shuttle *Atlantis*, *Magellan* left the cloud-streaked Blue Planet on May 4, 1989, for a destination 950 million miles away. On August 10, 1990, the craft entered orbit around Venus, where it began a radar mapping mission that has shown us Venus in detail that outstrips our knowledge of Earth itself, where large areas of sea basins remain uncharted.

miles and as far as 5,296 miles from the surface, *Magellan* completes one north-to-south orbit every three hours and fifteen minutes. In its imaging mode, the craft points its high-gain antenna toward the ground during the portion of its orbit when it is nearest the planet, allowing its radar system to record radar echoes from a strip about 16 miles wide and 10,000 miles long. When the spacecraft moves out into the distant part of its orbit, it turns its high-gain antenna toward Earth to transmit the data just collected, then turns again toward Venus to resume mapping. Each strip slightly overlaps the one before, allowing scientists to put the swaths together into image mosaics.

Readings from the first test mapping strip were very good, much to the relief of Project Scientist Stephen Saunders, who for more than a decade had been spearheading the group of nearly thirty scientists associated with the mission. With a resolution sharp enough to distinguish features roughly the size of a football field, *Magellan*'s imaging was some ten times better than the data returned by the Venera 15/16 missions.

For the next 243 days—the time it takes Venus to rotate once on its axis beneath the plane of *Magellan*'s orbit—the spacecraft performed valiantly, despite problems with its data tape recorders and another, albeit briefer, loss of signal in mid-November. By mid-May 1991, the end of its first mapping

Golubkina Crater. In the lower part of this composite image, data from the 1983 Venera 15/16 missions yields a fuzzy outline of the central peak and terraced walls of this twenty-mile-wide meteor crater. In the upper portion, *Magellan*'s radar, which can distinguish features as small as 400 feet, brings the crater into sharper focus.

River Styx. Shown here is part of a 4,200-mile-long channel nicknamed the River Styx. Averaging 1.25 miles in width, it sometimes winds uphill, a sign that the ground shifted after it was formed. What carved the channel is a mystery. Even under Venus's extreme surface temperature, most lava would not stay hot and fluid enough to flow so uniformly for so far.

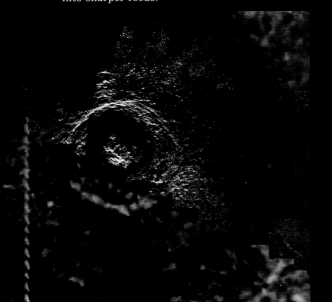

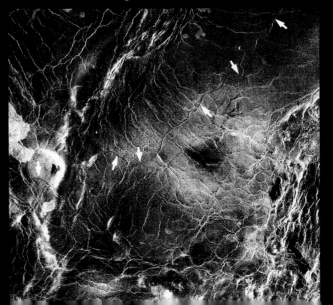

cycle, *Magellan* had surveyed about 90 percent of the Venusian globe. As the probe set out on its second mapping tour to fill in a variety of gaps, data-hungry scientists on Earth were poring over the material already in hand. The picture that emerged would prove riveting—and sometimes confounding.

VENUS: DEAD OR ALIVE?

In studying the radar readings, planetary scientists confirmed the impression of a youthful surface gained from previous probes. So far, *Magellan* has found no features older than one billion years and most of the surface is probably no more than about 400 million years old—younger by a few billion years than the average surface ages of the Moon, Mercury, and Mars.

Craters are, indeed, largely missing. In part this is because the planet's terrifically dense atmosphere acts as a shield against small- and medium-size meteoroids. But even the large chunks of space debris that should have made it to the surface seem to have left relatively few scars—erased, scientists are now agreed, by volcanic eruptions that spilled vast quantities of lava over hundreds or even thousands of square miles. According to Stephen Saunders, more than 80 percent of Venus is covered with volcanic plains material. With geological activity so palpably recent, project scientists were keen to find, in

Pancake Domes. About 2,500 feet high and fifteen miles wide, so-called pancake domes punctuate Venusian terrain. Formed when thick, pasty lava oozed through a central vent to the surface, the domes often show fractures that resulted as the lava cooled and cracked, as well as central pits that may have been created when lava ebbed away.

Alpha Regio. In the northern region of this highland, a 375-mile-long swatch of twisted terrain suggests the effects of crust-wrenching geologic forces. Such terrain scars about one-fifth of the globe, but the underlying tectonics remain a puzzle. Spokelike faults around an anomaly dubbed ''The Tick'' *(upper right)* may have been carved by lava.

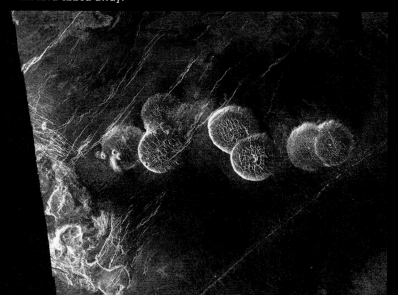

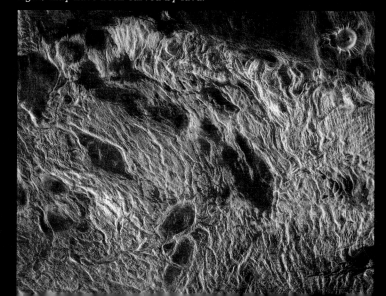

effect, the smoking volcano, or any other signs that the planet was alive and kicking internally. As *Magellan* began transmitting data from its second pass, the researchers looked for changes in the topography the spacecraft revisited.

One region that scientists were anxious to look at was the southern highland Aphrodite Terra, where, they hoped, they might get a clear indication of the nature of Venusian tectonics. Geologists James Head, a member of the Magellan science team, and Larry Crumpler, Head's colleague at Brown University, had predicted that the area would be a zone where the crust is separating and molten rock is welling up from the interior, spreading to form new crust as it does along Earth's Mid-Atlantic Ridge.

But *Magellan*'s data showed no sign of the features that would support the theory—no indication, for example, that the region's linear troughs align with a presumed spreading center, as do their terrestrial counterparts. Two schools of thought now suggest that the highland either stands over a rising hot spot in the mantle or that it might be a place where the crust is being subducted, or pulled back down into the mantle. *Magellan*'s results so far are inconclusive, although Saunders and his colleagues agree that there is convection in the mantle—a significant discovery in itself. For now, it appears that Earth-based models may not hold for Venusian tectonics.

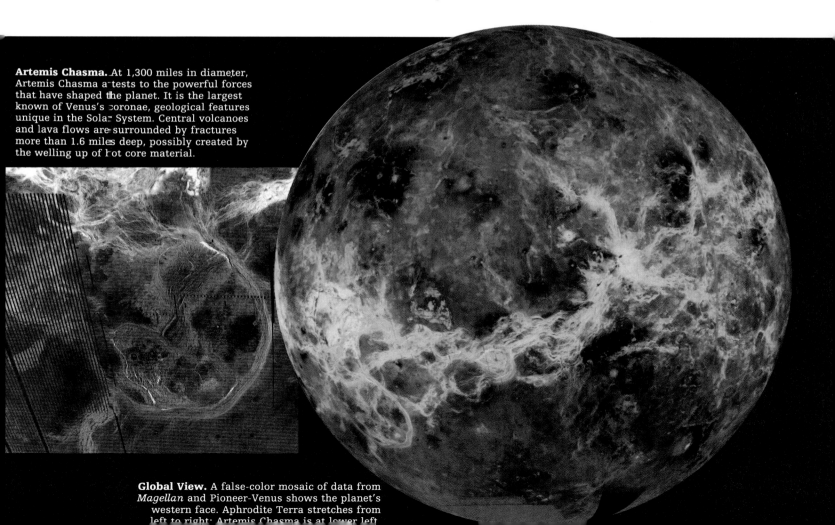

Artemis Chasma. At 1,300 miles in diameter, Artemis Chasma attests to the powerful forces that have shaped the planet. It is the largest known of Venus's coronae, geological features unique in the Solar System. Central volcanoes and lava flows are surrounded by fractures more than 1.6 miles deep, possibly created by the welling up of hot core material.

Global View. A false-color mosaic of data from *Magellan* and Pioneer-Venus shows the planet's western face. Aphrodite Terra stretches from left to right; Artemis Chasma is at lower left

PARTIAL ANSWERS, MORE QUESTIONS

As *Magellan* continued to send back its mapping reports, other geological features unique to Venus came into sharper focus. The circular oddities called coronae, first detected by the Venera 15/16 mission, were now seen to have volcanic peaks, some measuring as much as 200 miles across. Although scientists still do not have all the answers, it would appear that coronae are both volcanic and tectonic in origin.

Another enigma was a phenomenally long, narrow channel that has been nicknamed the River Styx. Although the Venera probes imaged part of it in the early 1980s, *Magellan* found that it runs for some 4,200 miles, making it the longest known channel in the Solar System. The great mystery is: What made it? *Magellan* has found absolutely no sign that Venus ever possessed running water on its surface, which leaves lava as a possibility. But to have carved a channel of such incredible length, lava would have to have been hot and fluid to a degree unknown on Earth. Scientists speculate that the river flowed with a very fluid form of lava, rich in carbon dioxide, or with sulfur, which could remain fluid on the Venusian surface. The River Styx is intriguing for another reason: Its meandering path attests to the planet's shifting crust. Rolling uphill as well as down, the channel in some places follows hills as much as a mile high—which presumably did not exist at the time the river ran.

By early January of 1992, as *Magellan* neared the end of its second mapping cycle, 95 percent of the planet had been surveyed. Further investigations, into the planet's gravity, were planned for later segments of the extended mission. But on January 4, the probe once more stopped transmitting. Reluctant to use a faulty backup transmitter, project engineers decided to turn off the spacecraft's tape recorders while the probe's handlers wrestled with the problem. Finally, since so much of the imaging mission was already accomplished, engineers settled for retrieving at least half of *Magellan*'s imaging and altimetry data by babying the backup transmitter along. Luckily, the slower transmission rate has no effect on the Venus gravity experiments.

Meanwhile, mission scientists have turned to the mountains of *Magellan* data still to be digested. The hardy probe has presented Earthbound researchers with answers to some long-standing questions while adding new puzzles of its own. For example, if Venus does not possess Earthlike plate tectonics, just how does it work? The leading theory is that convection in the planet's mantle, in effect, picks up and drags the overlying crust into new formations. Similarly, the probe revealed a planet that has been volcanically resurfaced to a degree matched or surpassed only by Jupiter's moon Io and by Earth itself. But Venusian volcanism has taken forms—such as the River Styx—that have no real counterpart here, and that may be related to the planet's tectonics in different ways. In the end, *Magellan* laid to rest any vestige of the romantic view of Venus as Earth's erstwhile twin. It is now clear that Venus does not and never did—at least not as recently as one billion years ago—possess planet-wide bodies of water or a more Earthlike climate. Its history and its destiny are, like Earth's, uniquely its own.

ANATOMY OF A
PLANETARY HELL

S wathed in a heavy veil of clouds, Venus long suc-
ceeded in hiding her true nature from astronomers
peering through earthly telescopes, giving rise to
fantasies of an otherworldly, tropical paradise that
might be suitable for terrestrial life. With the advent
of radar astronomy and robotic scouts, this romantic reputation was
ruined. In the 1970s, exploratory space probes investigated the planet's atmosphere and found the thick Venusian veil to be made primarily of corrosive sulfuric acid, scaldingly hot and swirling around
the planet at hundreds of miles per hour. The fiercest earthly hurri-
cane is a mere squall by comparison.

Not only is Venus's atmosphere chemically inimical, it is so dense
that it bears down on the planet's surface with a pressure equal to that
found at 3,000-foot depths in Earth's oceans. Moreover, it acts as a
heat trap, escalating temperatures until the surface is hot enough to
ignite paper instantly, even at night—a night that lasts four months
because of the planet's sluggish rotation.

As revealed in the following pages, the dynamics of Venus's atmosphere have created a world that looks more like the pit of hell
than the Garden of Eden.

ENCOUNTERING THE SOLAR WIND

The outermost layer of the Venusian atmosphere is a flexible envelope called the ionosphere, so named because it is made up of ionized, or charged, particles. Throughout this region, electrically neutral atoms and molecules of various gases are bombarded by ultraviolet rays from the Sun. The high-energy radiation knocks loose some negatively charged electrons to create positively charged atoms and molecules. (Electrons flowing to the dark side of the planet strike and ionize other gas particles to create and maintain the ionosphere even in the absence of solar radiation.)

The resulting plasma, as the particle soup is called, interacts with the solar wind, a thinner but hotter plasma streaming from the Sun at a speed of about 300 miles per second. Where the two plasmas meet, the interaction among the two sets of charged particles generates a complex magnetic layer that shields the planet from most of the effects of the solar wind. Fluctuations in the speed and density of the more powerful solar wind cause the ionosphere to billow and flap, shifting in altitude above the surface by as much as 500 miles.

Electromagnetic interaction with the solar wind compresses Venus's ionosphere on the sunward side and makes it bulge on the far side of the planet (below). The movement of charged particles at the top of the ionosphere, a region known as the ionopause, generates a magnetic field that deflects the solar wind and its magnetic field around the planet along a front called the bow shock. However, some solar wind particles penetrate the bow shock, reaching the ionopause. Yet to be explained are several interesting phenomena: lines of magnetic force called flux ropes that twine beneath the ionopause; 1,500-mile-wide holes in the ionosphere on the night side that plunge to within 100 miles of the surface; and clouds of plasma, perhaps torn from the ionosphere by the solar wind, that seem to drift into interplanetary space.

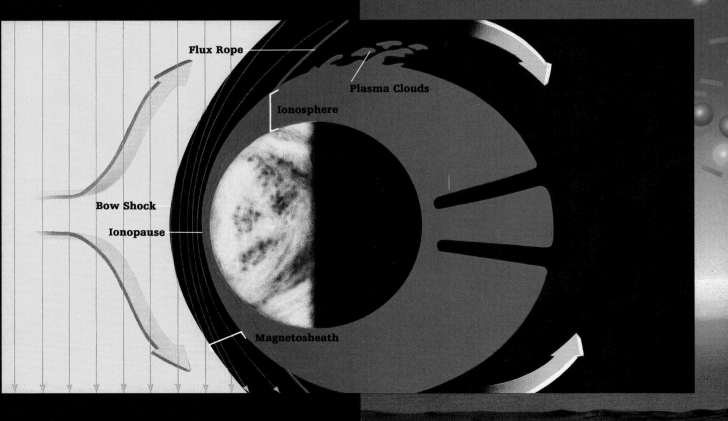

Flux Rope

Plasma Clouds

Ionosphere

Bow Shock

Ionopause

Magnetosheath

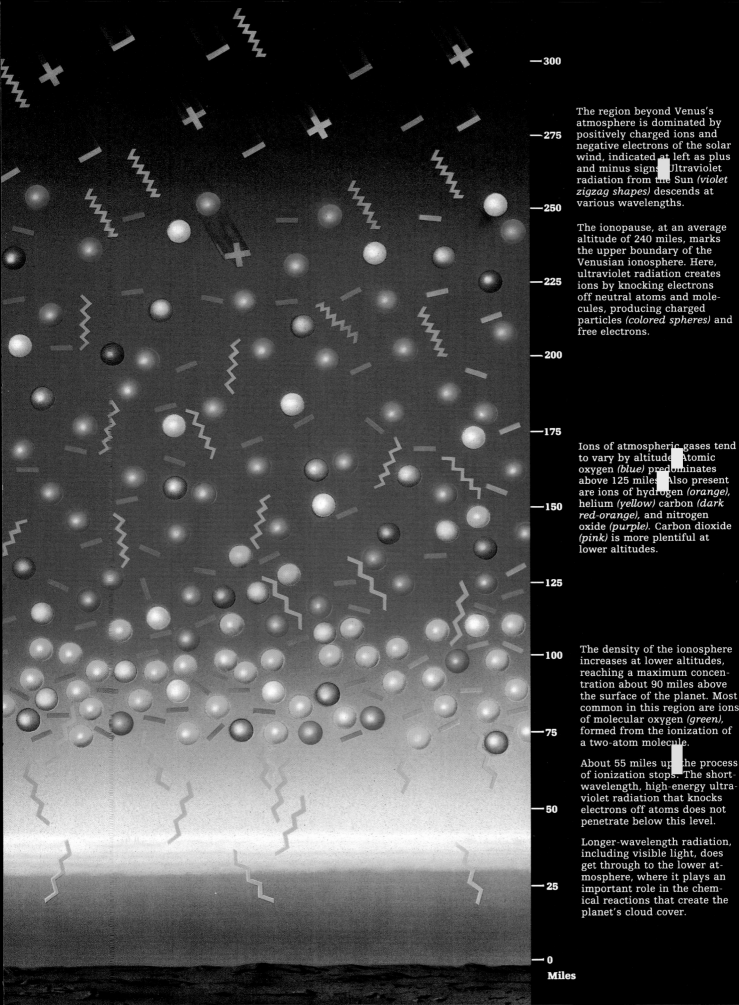

The region beyond Venus's atmosphere is dominated by positively charged ions and negative electrons of the solar wind, indicated at left as plus and minus signs. Ultraviolet radiation from the Sun *(violet zigzag shapes)* descends at various wavelengths.

The ionopause, at an average altitude of 240 miles, marks the upper boundary of the Venusian ionosphere. Here, ultraviolet radiation creates ions by knocking electrons off neutral atoms and molecules, producing charged particles *(colored spheres)* and free electrons.

Ions of atmospheric gases tend to vary by altitude. Atomic oxygen *(blue)* predominates above 125 miles. Also present are ions of hydrogen *(orange)*, helium *(yellow)* carbon *(dark red-orange),* and nitrogen oxide *(purple)*. Carbon dioxide *(pink)* is more plentiful at lower altitudes.

The density of the ionosphere increases at lower altitudes, reaching a maximum concentration about 90 miles above the surface of the planet. Most common in this region are ions of molecular oxygen *(green)*, formed from the ionization of a two-atom molecule.

About 55 miles up the process of ionization stops. The short-wavelength, high-energy ultraviolet radiation that knocks electrons off atoms does not penetrate below this level.

Longer-wavelength radiation, including visible light, does get through to the lower atmosphere, where it plays an important role in the chemical reactions that create the planet's cloud cover.

—300

—275

—250

—225

—200

—175

—150

—125

—100

—75

—50

—25

—0

Miles

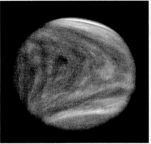

THE CLOUD COVER: A HIGH-SPEED DANCE

As viewed from Earth, the most salient characteristic of Venus is its dense swaddling of clouds, a twelve-mile-thick layer beginning just below the ionosphere, about forty-five miles above the planet's surface. Variations in solar warming, from night side to day and from pole to equator, determine the patterns of movement in this layer of the atmosphere.

At the top of the cloud layer, the slight temperature differential between the night and day sides of the planet causes a difference in atmospheric pressure. Masses of gas streaming from high-pressure areas to low become 250-mile-per-hour winds roaring from east to west. While Venus itself rotates at a snail's pace, taking 243 days to complete one spin on its axis, this top layer of clouds whips all the way around the planet in just four days.

The cloud-top winds spiral slightly toward the poles, as a consequence of temperature differences arising from the different angles at which solar radiation strikes the atmosphere. Clouds at the equator, where the Sun is more or less overhead, are heated more than those at the poles, where the warming rays are more oblique. Within the cloud layer, this north-south temperature gradient gives rise to a so-called Hadley cell pattern *(right)*, named for the eighteenth-century English physicist who described a similar terrestrial phenomenon. At this level, the weight of Venus's massive atmosphere slows the Hadley pattern winds to a sedate twenty miles per hour.

A five-day sequence of ultraviolet photographs reveals long-term features of Venus's cloud cover. Rotating at speeds sixty times as fast as the planet itself, cloud tops form perceptible bands around the north and the south poles. The dark horizontal Y pattern seen in the first, fourth, and fifth photos is a semipermanent feature that can be tracked for several days as the clouds travel around the planet.

Beneath the high-speed winds at the top of Venus's clouds, scientists theorize, are various layers of circulation. Immediately below the cloud tops, temperature differences generate movement at right angles to the east-to-west winds. Clouds above the equator, heated more quickly than those over polar regions, rise and then flow toward the poles. There they cool and sink, then flow back toward the equator. In the next layer down, gas dragged along by friction with the moving clouds rotates in the opposite direction. Some scientists believe another layer exists lower still; also subject to friction, it reverses direction yet again. Interaction of the wind patterns produces a swirling tornado-like vortex at each of the planet's poles.

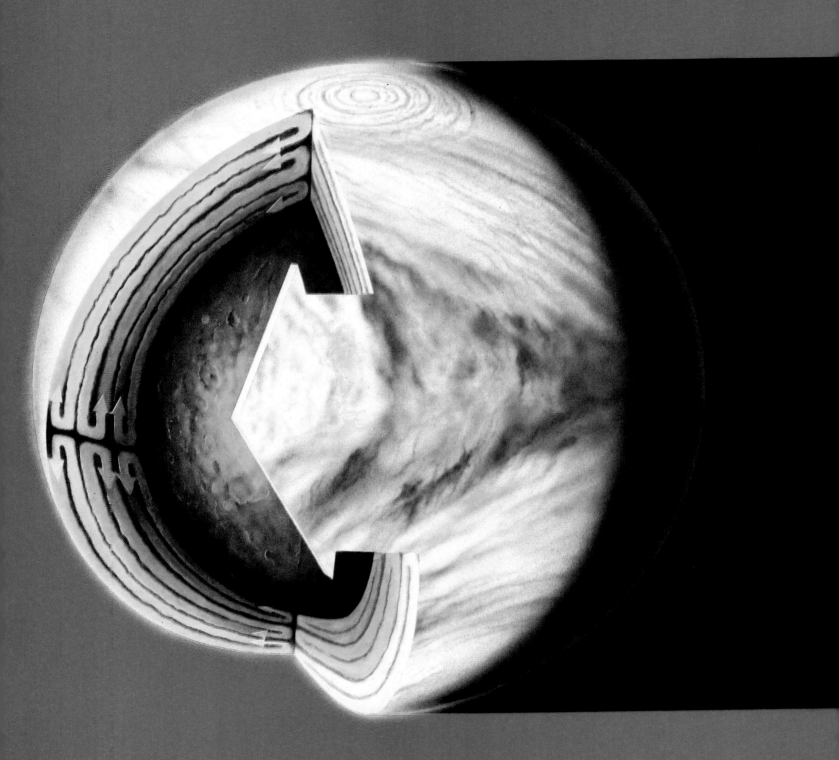

An Inferno of Trapped Sunlight

Venus owes its ovenlike climate to two chemical compounds: carbon dioxide, which makes up about 96 percent of the atmosphere, and sulfur, which accounts for only a minute fraction but is concentrated in the planet's sulfuric acid clouds. Transparent to visible light but an efficient absorber of infrared, carbon dioxide plays a key role in the greenhouse effect, which allows the hottest planet in the Solar System to build up its enormous surface temperature from the solar radiation that pierces the clouds.

The cloud tops reflect most of the solar radiation that hits them, and the clouds themselves soak up more than two-thirds of the remainder; only seven percent of the solar energy gets past the cloud barrier, and only two to three percent reaches the ground. The most opaque part of the cloud layer is at the bottom, where sulfuric acid droplets are concentrated with a density of 400 per cubic centimeter—similar to a heavy smog on Earth. Density drops fourfold in the middle of the clouds but increases to a thin smog in the uppermost layer.

A downpour from these clouds would suffuse the planet's lower atmosphere with harsh, corrosive liquid, but because the lower level lacks the vigorous movements of hot and cold masses of gas that influence Earth's weather, the clouds remain relatively stable, giving off no more than a light acidic drizzle.

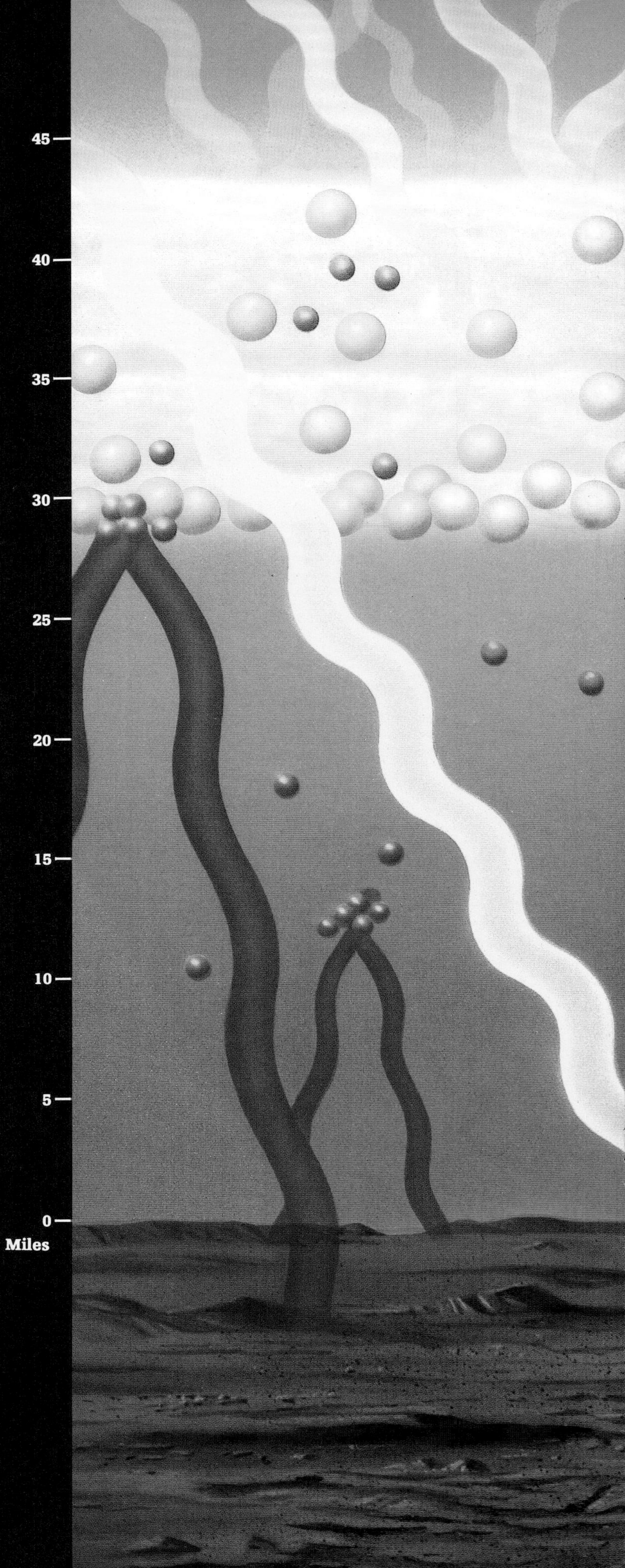

The so-called greenhouse effect heats Venus with the small amount of sunlight *(yellow)* that penetrates the clouds. Warmed by this solar energy, the surface emits infrared radiation *(red)* back into the atmosphere. Carbon dioxide at and below the level of the clouds absorbs the infrared and reradiates it, maintaining the heating cycle that keeps the planet's surface temperature at over 800 degrees Fahrenheit— hot enough to melt some metals.

45 —

40 —

35 —

30 —

25 —

20 —

15 —

10 —

5 —

0 —

Miles

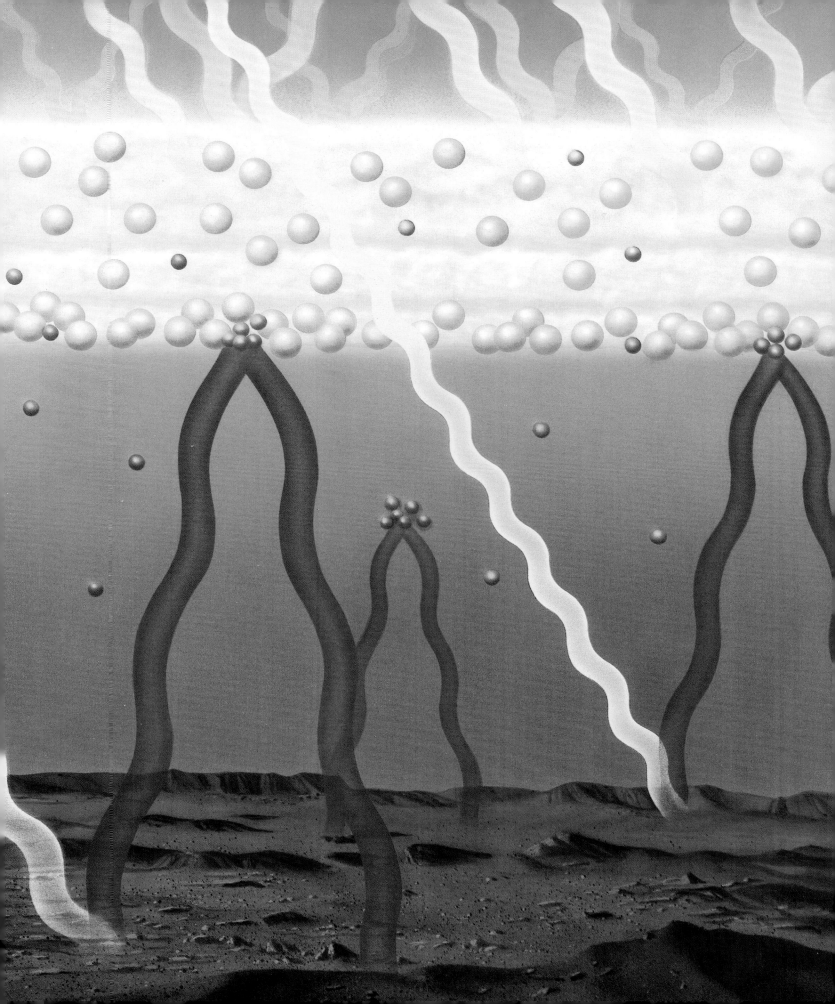

THE SAVAGE
SULFUR CYCLE

Unlike the clouds of Earth, which form by simple evaporation and condensation, Venusian clouds are believed to result from a complex sequence of volcanic activity, solar radiation, and multiple chemical reactions. One widely held theory describes a process that involves not only the atmosphere but also the surface of the planet, its interior, and the ultraviolet radiation that reaches Venus from the Sun.

The events that produce the clouds can be understood as three interlinked cycles—geologic, slow atmospheric, and fast atmospheric *(right)*. In the geologic phase, volcanic eruptions or lava flows bring up sulfur from the bowels of the planet. The atmospheric cycles are characterized by two different types of chemical reactions. Up to an altitude of about twenty miles, heat-driven thermochemical processes predominate, and sulfur compounds are transformed in the pressure cooker of the atmosphere. Above that altitude, ultraviolet radiation breaks complex molecules into smaller molecules or atoms. These recombine in turn to form new compounds, which may be separated once again in the complex chemical dance of the sulfur cycle.

A single molecule takes about a year to pass through the high-altitude fast atmospheric cycle and a decade to complete the lower, slower cycle. In the geologic phase, a molecule may take as long as two million years to finish its rounds.

45 —

40 —

35 —

30 —

25 —

20 —

15 —

10 —

5 —

0 —

Miles

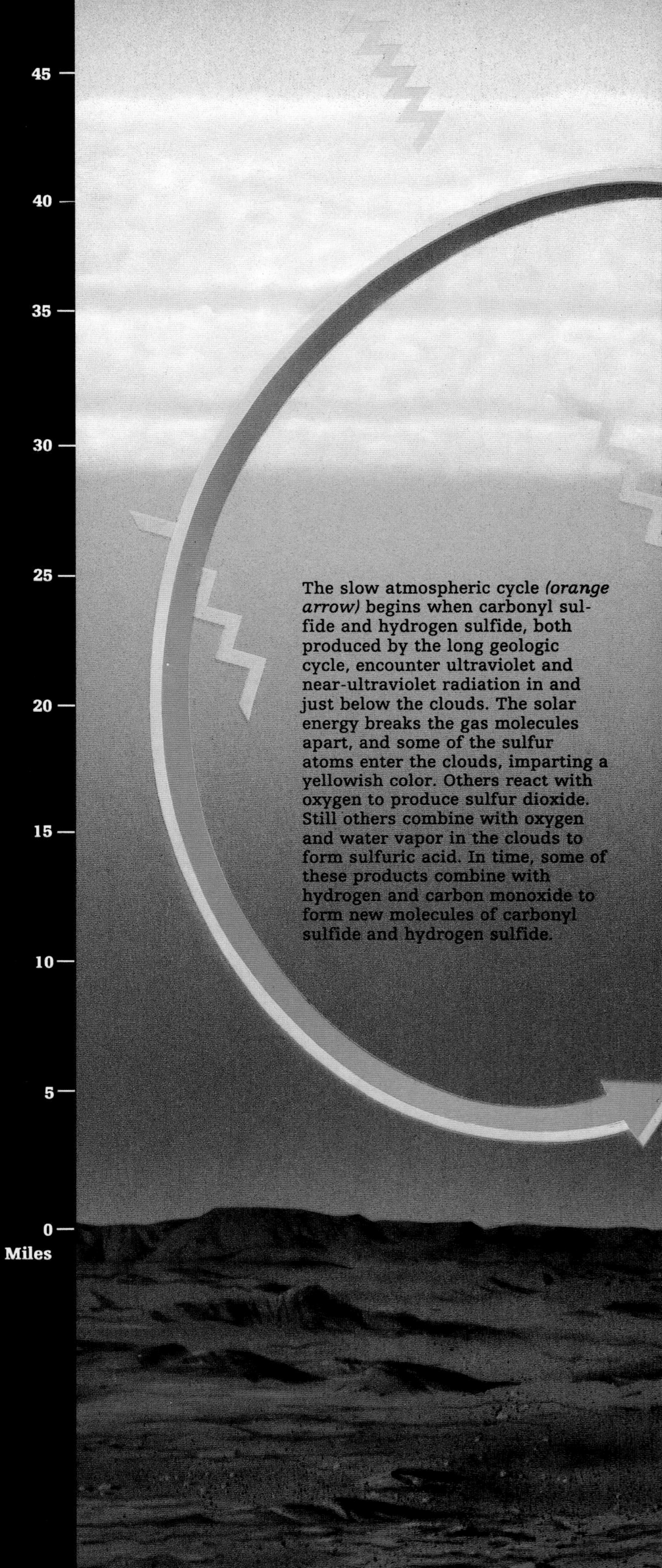

The slow atmospheric cycle *(orange arrow)* begins when carbonyl sulfide and hydrogen sulfide, both produced by the long geologic cycle, encounter ultraviolet and near-ultraviolet radiation in and just below the clouds. The solar energy breaks the gas molecules apart, and some of the sulfur atoms enter the clouds, imparting a yellowish color. Others react with oxygen to produce sulfur dioxide. Still others combine with oxygen and water vapor in the clouds to form sulfuric acid. In time, some of these products combine with hydrogen and carbon monoxide to form new molecules of carbonyl sulfide and hydrogen sulfide.

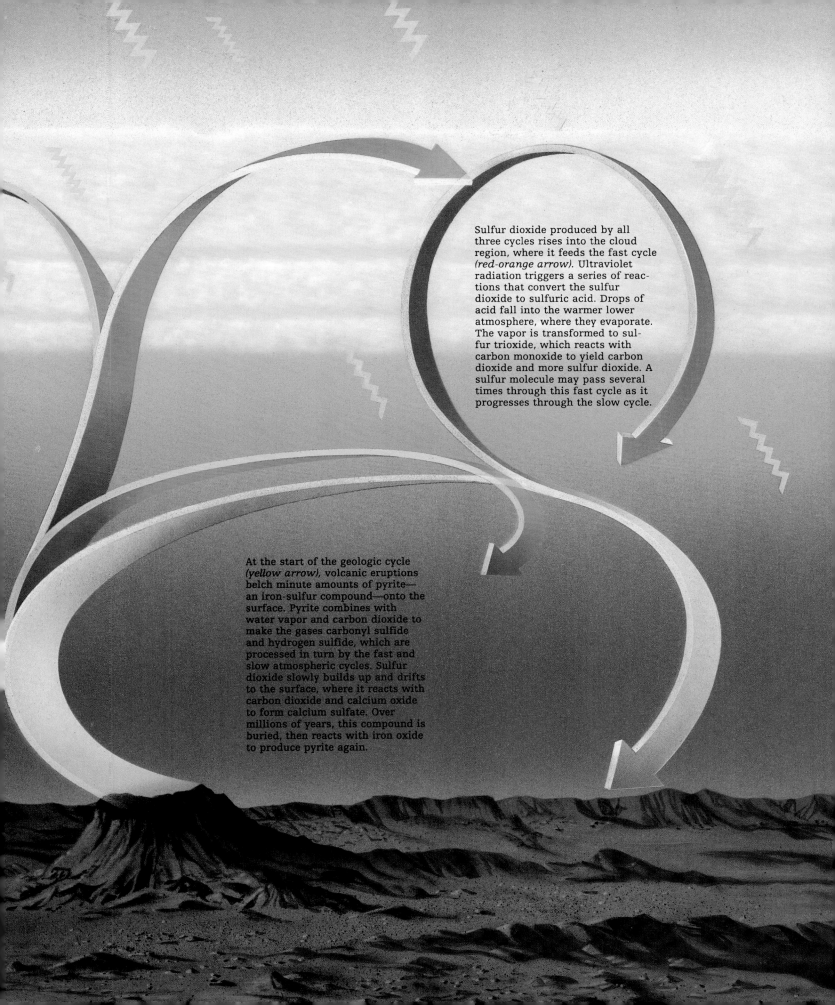

Sulfur dioxide produced by all three cycles rises into the cloud region, where it feeds the fast cycle *(red-orange arrow)*. Ultraviolet radiation triggers a series of reactions that convert the sulfur dioxide to sulfuric acid. Drops of acid fall into the warmer lower atmosphere, where they evaporate. The vapor is transformed to sulfur trioxide, which reacts with carbon monoxide to yield carbon dioxide and more sulfur dioxide. A sulfur molecule may pass several times through this fast cycle as it progresses through the slow cycle.

At the start of the geologic cycle *(yellow arrow)*, volcanic eruptions belch minute amounts of pyrite—an iron-sulfur compound—onto the surface. Pyrite combines with water vapor and carbon dioxide to make the gases carbonyl sulfide and hydrogen sulfide, which are processed in turn by the fast and slow atmospheric cycles. Sulfur dioxide slowly builds up and drifts to the surface, where it reacts with carbon dioxide and calcium oxide to form calcium sulfate. Over millions of years, this compound is buried, then reacts with iron oxide to produce pyrite again.

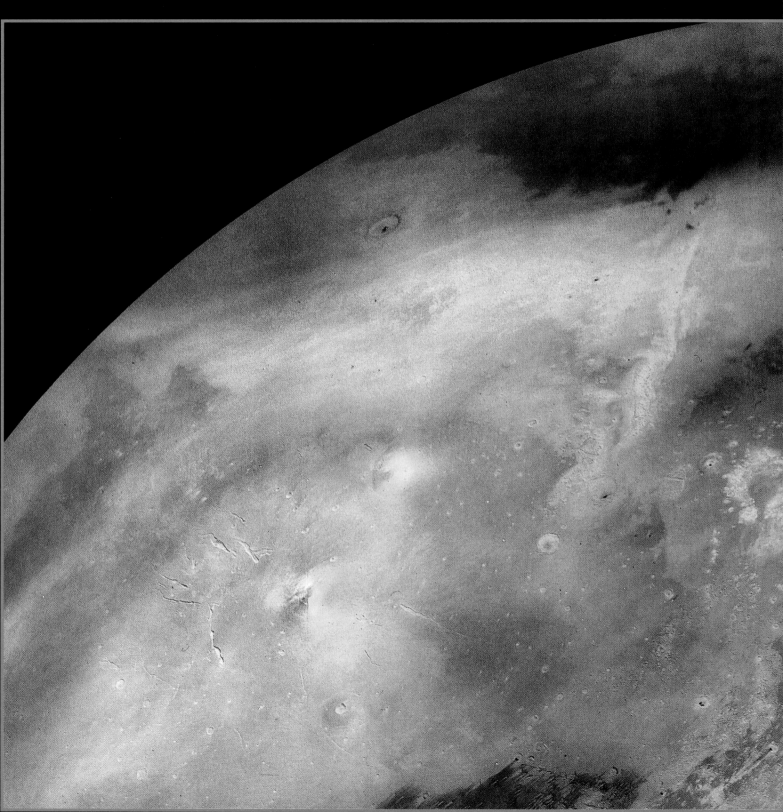

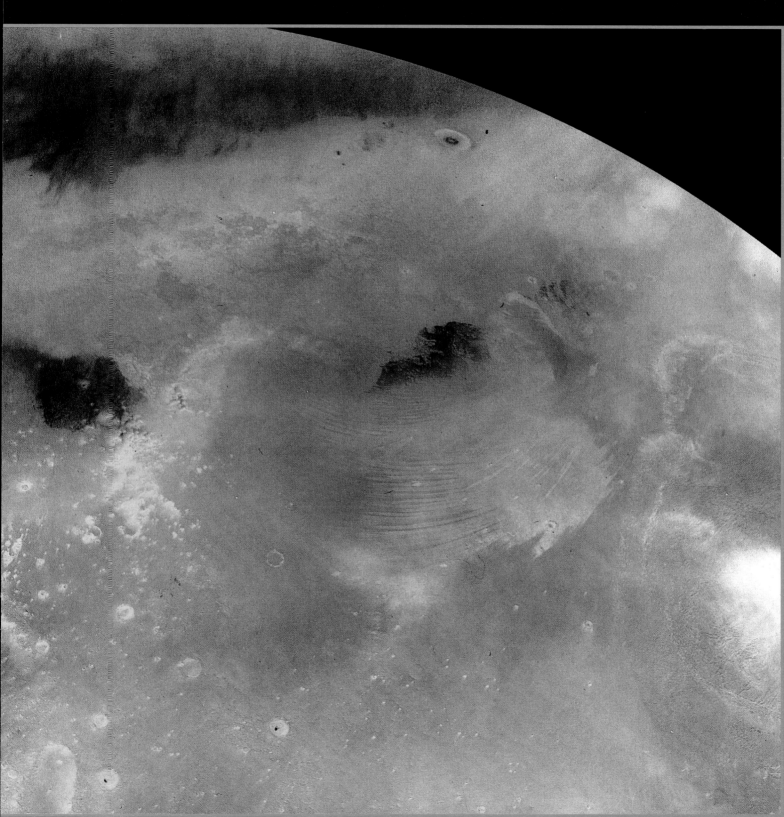

Like islands in a red sea, dark rocky areas stripped of dust by brutal Martian windstorms poke through the rusty blanket that gives the Red Planet its name.

ircling Mars at an altitude of 1,025 miles, *Mariner 9* looked down and saw virtually nothing. A month before, on November 13, 1971, the spacecraft had become the first man-made object ever to orbit another planet. Its cameras were capable of mapping the Martian surface from pole to pole, picking out details as small as a football field. But the cameras were practically useless: Mars had welcomed its new companion by shrouding itself in a colossal dust storm. Pictures of the planet, showing great, rusty billows whipped high into the rarefied atmosphere by 100-mile-per-hour winds, looked like closeups of a tennis ball.

This stroke of bad luck threw observers at the Jet Propulsion Laboratory into near despair. Many saw years of their work coming to naught. The blow was softened, though, by the common belief that Mars actually had little to hide. The space probe's predecessors in the 1960s—flyby missions, not orbiters—had revealed an apparently barren world, one that was pockmarked with craters and as geologically dead as the Moon. Its unremarkable surface seemed to lack even the relief of the Moon's jagged mountains. Now, however, scientists poring over *Mariner 9*'s pictures were perplexed by the only permanent features visible to the orbiting spacecraft: four dark spots that showed through the storm, even at its wildest. They could only be mountains. But if they were, then Mars was a far different place than expected.

In mid-December, the winds abated slightly, and evidence of that difference began to emerge. As the dust clouds settled, the spots took shape in the sunlight. Image by image, their true identity became clear: They were not just mountain peaks but the crests of volcanoes—volcanoes so huge they dwarfed any on Earth or in the known Solar System. More signs of geologic activity on a grand scale appeared. *Mariner 9*'s cameras photographed a vast rift valley, two miles deep and a hundred miles wide, slashing across the Martian surface for 3,000 miles.

These discoveries were tremendously exciting to planetary scientists, but they paled before Mars's final revelation. As the last of the dust subsided, *Mariner 9*'s cameras showed that the Martian plains were laced with thousands upon thousands of channels—sinuous gorges that looked like nothing so much as the water-carved canyons and arroyos of the

American Southwest. Mars, now a world that is more arid than any terrestrial desert, had once seen the fluid of life course freely over its surface.

AN ANCIENT MYSTIQUE
It was not the first time that Mars had surprised Earth-bound observers, nor would it be the last. The shimmering, yellow-brown disk had fascinated humans for hundreds of years. Astronomers had focused increasingly sophisticated instruments on the planet, but in the three centuries preceding the dawn of planetary exploration they had managed to gather only a handful of trustworthy facts. Mars was half the size of Earth, with a tenth the mass and about a third the surface gravity. Orbiting more than 140 million miles from the Sun, the planet had a remarkably Earth-like rotation period of twenty-four hours, thirty-seven and a half minutes, and turned on an axis that was tilted twenty-four degrees to its orbital plane—just a degree more than Earth. Because of its greater distance from the Sun, however, its year was 687 Earth days long. The surface bore markings that varied with the seasons, including brilliant white polar caps. A thin, cold atmosphere supported clouds and occasional dust storms. Finally, two tiny moons, Phobos and Deimos, circled Mars at respective distances of 5,860 miles and 14,690 miles.

But the mystique of Mars had little to do with mere fact. To many people, the planet was a place of dreams and haunting mystery. The Mars of the English writer H. G. Wells was a dying world populated by a desperate race. His 1898 novel of interplanetary invasion, *War of the Worlds,* ensured that generations of readers would hear the name of Mars with a tiny shiver of fear. Edgar Rice Burroughs, the American creator of Tarzan, wrote eleven best-selling novels about the planet, called Barsoom by its inhabitants. Readers could not get enough of Burroughs's Martians, who came in assorted colors and forms and included such characters as the beautiful Princess of Helium, Dejah Thoris.

All these works shaped the popular conception of Mars, their influence extending to those who would grow up to be scientists. Cornell astronomer Carl Sagan, a *Mariner 9* science team member, recalled that as an eight-year-old boy he spent hours in an empty field with his arms outstretched, trying to wish his way to Mars like Burroughs's hero John Carter. Ray Bradbury, who was a science-fiction writer himself, nominated Burroughs as the true father of the space age, because he fostered dreams of traveling to the distant planet.

Although few scientists believed in Martian canals or ancient civilizations, the ideas still resonated. Dry and cold as it might be, Mars remained the most plausible home for life in the Solar System outside Earth itself. More than any other world, it seemed to promise an answer to an age-old question: Are we alone?

The call of Mars was so strong that probes were launched almost as soon as suitable rockets became available. The first, the Soviet Union's *Mars 1,* was sent on its way on November 1, 1962, barely five years after Sputnik. The one-ton spacecraft was a testament to the power of Soviet boosters, but it began a series of humiliating failures for the Soviet Mars program: Ground controllers lost contact with *Mars 1* in March 1963, months before it was due to reach its target.

The first U.S. attempt did no better. *Mariner 3* blasted off from Cape Canaveral aboard an Atlas-Agena rocket on November 5, 1964, just a little more than two years after the *Mars 1* launch, when the launch window for an economical trip to Mars opened again for a few weeks *(pages 35-43)*. A heat shield, designed to protect *Mariner 3* during its trip up through Earth's atmosphere, failed to release afterward, leaving the spacecraft trapped in a metal cocoon. Unable to extend its solar panels, *Mariner 3* ran out of power for its instruments and radios, which went dead soon after launch.

Fortunately, NASA had a second spacecraft, a twin to *Mariner 3,* waiting in the wings. On November 28, 1964, *Mariner 4* began an eight-month flight that brought it within 6,100 miles of the Red Planet's surface the following summer, on July 15. During the twenty-minute period when the probe's television camera had Mars in view, *Mariner 4* took twenty-one pictures of the southern hemisphere. Because the communications system needed more than eight hours to transmit a single image, the eagerly awaited pictures had to be stored on magnetic tape. *Mariner 4's* vision of Mars took ten days to unfold.

The formidable Stickney crater dominates one end of seventeen-mile-long Phobos *(above),* the innermost of the two Martian moons. Phobos orbits the planet three times each Martian day, traveling so fast that tidal forces—a kind of gravitational friction between the moon and the planet—cause the satellite to spiral slowly inward. Mars and Phobos are expected to collide in approximately 30 million years.

AN END TO SPECULATION

The first picture to arrive at JPL, showing a blurred arc of the planet against a dark sky, sent a ripple of excitement through a rapt audience of scientists and journalists. But the first closeup of Mars from space revealed next to nothing about the planet, and the next few pictures were equally uninformative: The probe was scanning a part of the planet where the Sun was high in the sky, so there were few shadows to highlight surface features. Nevertheless, scientists continued to anticipate that some sign of primitive life forms might be discovered.

That hope began to sink two days later with the arrival of frame seven. The field of view now included regions where it was afternoon, and shadows began to reveal the nature of the surface, with detail thirty times better than the best images made from Earth. What the pictures showed were craters, similar to the ones existing on the Moon. By frame eleven, which took in a seventy-five-mile-wide impact basin, the truth had become inescapable: Not only were these parts of Mars devoid of canals or of any other large-scale engineering works, they presented no sign of even the most rudimentary life.

Up close, dark areas that might have been vegetation proved equally disappointing, full of craters and other features that were no different from those in the light-colored regions. The dark areas did not even have sharply defined boundaries; the surface simply changed color gradually. By the time the sequence ended (the last few frames showed nothing but the darkness on the night side of Mars), whole centuries of hypothesis, guesswork, and wishful thinking had been swept away.

Readings from *Mariner 4*'s instruments drove more nails into the coffin of Martian fantasies. The air was as thin as Earth's at an altitude of fifteen miles; it would be unbreathable even if it were pure oxygen—which it was not. In contrast to Earth's gaseous envelope of nitrogen and oxygen, the Martian atmosphere was largely carbon dioxide. With surface pressure less than one percent that of the Earth at sea level, there could be no liquid water, because under such conditions the stuff would evaporate instantly. Mars was a desert. The *New York Times* caught the general mood of the astronomical community with the title of an editorial about *Mariner 4:* "The Dead Planet."

DIEHARD HOPES

The probe's revelations did not curb curiosity about Mars, however. There was still much to be learned about the planet's geology and atmosphere. In any case, *Mariner 4* had taken pictures of only one percent of the surface, and at a level of resolution that could easily have missed evidence of life on Earth. In the minds of some scientists—and many in society at large—the question of life on Mars remained unresolved. If there were no canal builders, there might yet be other forms of life that had adapted to the harsh Martian environment. Further exploration seemed only natural.

The next two missions to Mars, staged two launch opportunities later, were also flybys: *Mariner 6* came within 2,175 miles of the surface on July 31, 1969, as did *Mariner 7* five days later. With much-improved communications systems, these two spacecraft could send data almost 2,000 times faster than *Mariner 4;* as a consequence, they returned many more pictures, and these were of much better quality.

The images again showed mostly cratered terrain. More clearly than ever, they demolished any remaining notions of canals; there simply were none. Instrument readings buttressed previous findings about the thin, cold atmosphere of carbon dioxide. Other experiments showed that, unlike its neighbor Earth, Mars had no detectable magnetic fields or radiation belts. As the two spacecraft followed *Mariner 4* into permanent orbit around the Sun, they left a Mars that seemed a dull, geologically inert world, one little different from the Moon.

GOING BACK FOR MORE

It was in this context that the United States and the Soviet Union prepared new missions for the next launch opportunity, in May 1971. Their target was glaringly conspicuous in Earth's skies, making its closest approach since

With high-altitude clouds forming a thin line on the horizon, winter days offer a pristine view of Argyre basin in Mars's southern hemisphere. The approach of spring and warmer temperatures will bring a dense haze of condensed carbon dioxide, water vapor, and dust.

1924. In the early days of the space program, optimists had hoped that 1971 would be the year of the first manned expedition to Mars. In the event, the ventures were far less ambitious.

The U.S. effort was planned as a two-spacecraft mission, but bad luck intervened again: *Mariner 8* crashed into the Atlantic during launch, leaving its twin, *Mariner 9,* to carry on alone. Both of the Soviet vehicles, *Mars 2* and *Mars 3,* made the trip successfully.

Unlike their predecessors, all of these probes were designed to orbit Mars rather than just fly by. *Mariner 9* weighed more than a ton—nearly four times as much as *Mariner 4.* Almost half the weight was fuel required to slow the vehicle when it reached Mars so it could drop into orbit around the planet. The spacecraft had a miniature, on-board computer that allowed mission controllers at JPL to update its instructions and thus adapt the mission to changing circumstances—which was fortunate, since the probe would now have to carry out *Mariner 8's* tasks as well as its own. Among the instruments for the six scientific experiments on the probe were two cameras, each equipped with both wide-angle and telephoto lenses.

On November 14, 1971, a fifteen-minute firing of *Mariner 9's* rocket engine placed it in a twelve-hour orbit about 860 miles above the surface of Mars. On November 27 and December 2, it was joined there by the two Soviet orbiters. At 10,250 pounds apiece, these were the heaviest vehicles yet to visit the Red Planet. Shortly before going into orbit, *Mars 2* ejected a capsule containing a Soviet pennant. Though useless in scientific terms, it was the first man-made object to reach the surface.

In addition to its scientific instruments, *Mars 3* carried a landing craft with a camera. Although the dust storm was still raging, the lander was released shortly before the parent ship achieved orbit. Weathering the effects of hundred-mile-per-hour winds during its descent, it touched down safely near the southern polar cap on December 2. After a minute and a half, the vehicle began to transmit its first picture. For twenty seconds, all seemed well, although no details were discernible. Then all transmissions suddenly ceased. Soviet scientists later suggested that the lander's parachute, which had been jettisoned 100 feet above the ground, had drifted down on top of the probe.

A VISUAL FEAST

By this time, *Mariner 9* was already beginning its survey from orbit. For more than ten months, until it finally fell silent at the end of October 1972, the orbiter produced a flood of data. Scientists learned a lot about the atmosphere—its component gases and suspended particles, wind speeds, and the escape of molecules into space. They also got information about the planet itself, such as surface composition, temperature variations from day to night, the gravitational field, and the distribution of mass beneath the surface.

But what gripped the attention of scientists and public alike was the steady, astonishing flow of images, 7,329 in all. *Mariner 9* mapped virtually the entire planet to a resolution of less than one mile. For two percent of the surface, its cameras zoomed in tighter still, detecting features 100 to 300 yards across. These images revealed that Mars is geologically schizophrenic, with a transition between "southern" and "northern" terrain occurring along a line inclined roughly fifty degrees to the equator. The southern hemisphere is far more heavily cratered than the north, and it displays far less evidence of geologic activity. *Mariner 9*'s predecessors had shown such a Moon-like planet because, purely by coincidence, they had all taken their photographs in the southern hemisphere.

In the south, the multiplicity of craters told Mariner scientists that the surface was very stable and very old. The craters appeared to be relics of Mars's battered youth. Well before the voyage of *Mariner 9,* astronomers had become convinced that all the terrestrial planets had once been pummeled by rocks, boulders, asteroids, and comets—junk left over from the formation of the planets about 4.6 billion years ago. On Earth, erosion and the drifting of the continents have long since erased any trace of this era, but the craters and impact basins on the Moon bore eloquent witness. *Mariner 4* proved that the bombardment had extended out to Mars, and *Mariner 9* showed just how fierce it had been. In the southern hemisphere, the spacecraft's cameras found several large, basinlike areas, many surrounded by concentric rings of mountains. Enormous craters such as Hellas, nearly 1,200 miles wide, and Argyre, 560 miles in diameter, were almost certainly excavated by the impacts of giant asteroids.

A COLOSSAL QUARTET

In the north, by contrast, *Mariner 9* found a much younger surface. The northern hemisphere was essentially one vast lava flow; craters that might have existed there had been obliterated by volcanic activity on a colossal scale. The rare northern craters, with their fresh, sharp edges, had probably resulted from relatively recent impacts. The dominant features of this part of the planet were the four volcanoes that *Mariner 9* was able to make out above the dust storm, the largest of which lay at a point that had been observed through telescopes. Named Nix Olympica, the Snows of Olympus, because it often appeared to be a bright ring, the spot proved Olympic indeed: a great, sprawling mass rising some fourteen miles above the surrounding plain, nearly two and a half times the height of Everest above sea level. Its base was broad enough to cover most of Spain, and its edge was marked by a cliff face more than a mile high.

The other three volcanoes—dubbed North Spot, Middle Spot, and South Spot by *Mariner 9* scientists—were small only by comparison. Located to the southeast of Nix Olympica (soon renamed Olympus Mons), they were strung along a thousand-mile line just north of the Martian equator, across a region of concentrated volcanism named Tharsis. Measurements from *Mariner 9* indicated that Tharsis was a huge bulge, protruding more than four miles above the average surface level of the planet. Whatever internal forces had pushed it up seemed to have cracked the surface of Mars for thousands of miles in every direction. The 3,000-mile rift, dubbed Valles Marineris in honor of its robot discoverer, was the largest concentration of such cracks, a place where the surface of the bulge had split open.

WHENCE THE WATER?

The planet's most enigmatic attributes, the eroded channels, were concentrated in the equatorial region, viewed for the first time by *Mariner 9.* It was a wilderness of ravines, meanders, braided streambeds, and tributaries, a topography that instantly made water a critical issue in the study of Mars. If water carved the channels, even billions of years ago, scientists had to figure out where it had gone. The Martian atmosphere was not the answer: If all the atmospheric water were deposited on the surface, it would scarcely make a layer of dew. Nor were the polar caps adequate as reservoirs: Most scientists believed they held a substantial amount of frozen carbon dioxide, and even if they were entirely water ice, they held only enough to form a planetwide lake perhaps thirty feet deep. To create the channels, all this water would have to have been concentrated near the equator and somehow all released at once in a huge flash flood, an unlikely occurrence. Some scientists speculated that a vast amount of unseen water is perennially frozen into the Martian subsoil, a phenomenon known on Earth as permafrost.

The water debate emphasized the need to look not just at individual features of Mars but at the whole planet and its interacting systems. Any explanation of the channels—what had caused them and when—had to make sense in terms of the present-day atmosphere, which governs the temperature changes that cause melting and freezing. Temperature is also affected by dust. The dust, in turn, could be a product of many things, such as volcanism and erosion by wind and water. Mars held challenges for geologists, chemists, meteorologists, physicists, and perhaps even biologists. In transforming Mars into a real place, with real physical attributes, the Mariner probes dramatically broadened the scope of planetary science.

The first strokes in the emerging picture of Mars were painted by geologists, who could refer to well-studied parallels on Earth. The stupendous mountains that crowned the Tharsis Bulge strongly resembled the volcanoes of Hawaii, where a hot spot—a small region where molten rock from deep below the surface burns its way through the crust—sends out a thin, runny kind of lava that hardens into basalt. In both places, the mountains are broad, domelike structures, pushed up by lava accumulating under the surface; large, craterlike summit calderas mark places where lava chambers have drained and collapsed. The sheer size of the Martian volcanoes was a testament to how long and how steadily the subsurface forces had been working: The cones had apparently spewed lava and built themselves up for millions upon millions of years.

Some geophysicists noted that Tharsis, with its volcanoes and its chasms like Valles Marineris, seemed similar to systems of seafloor ridges, hot spots,

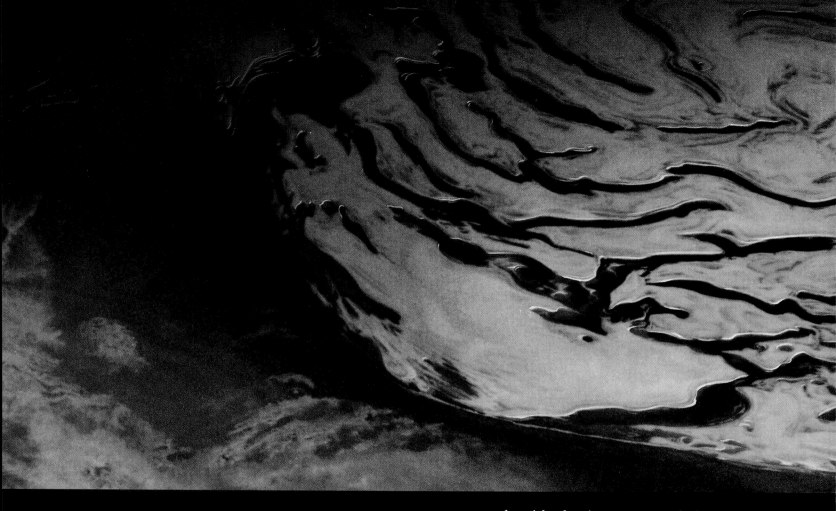

A POLAR CYCLE

In a global game of atmospheric ping-pong with no terrestrial counterpart, the annual growth and shrinkage of the polar caps on Mars may generate powerful cyclical winds that affect the whole planet. The Martian atmosphere—about 95 percent carbon dioxide and less than a hundredth as dense as Earth's—

responds with alacrity to seasonal changes in temperature. As the north pole warms in spring and summer, water ice in the polar cap *(above)* quickly begins to evaporate and mix with atmospheric carbon dioxide. The winter cap, which extends south to middle latitudes, shrinks around the edges, to the boundaries of the residual, or permanent, cap.

At the south pole, meanwhile, winter approaches. As temperatures sink to minus 190 degrees Fahrenheit,

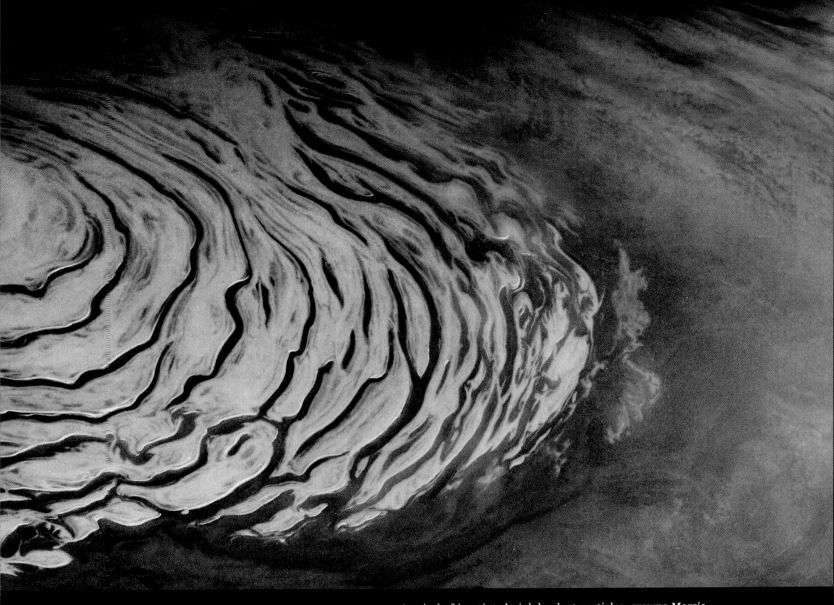

carbon dioxide begins to freeze out of the atmosphere onto the surface of the south polar cap. Atmospheric pressure drops by some 30 percent across the planet, producing a rush of air from north to south at high altitudes. When the seasons switch and the carbon dioxide freeze-out occurs over the north pole, the atmospheric rush is reversed. These high-altitude shifts trigger events that can explode into planet-shrouding dust storms *(pages 116-117)*.

A spiral of ice, tinted pink by dust particles, crowns Mars's north pole with a permanent, or residual, cap. In late spring, the pole basks in perpetual sunlight, and as the polar cap melts around the edges, evaporated water ice forms a hazy hood in the atmosphere over the region.

A WIND-BORNE PALL

During the southern spring, a season of rapid and intense heating, a strong atmospheric flow begins to transport dense, cold air from the south pole of the planet toward regions of lighter, warmer air nearer the equator. Passing over downslopes of precipitous terrain with peaks two to three times as high as Earth's

Mount Everest, winds can accelerate to hundreds of miles per hour.

At these speeds, the thin air of Mars can pick up fine dust grains, carrying them high into the atmosphere on vertical currents. Cells of circulation that exchange warm and cold air between equatorial regions and the poles carry the dust northward. As more particles are borne aloft, they absorb solar radiation, heating the atmosphere and further strength-

ening the winds. More wind rips more soil from the surface into the atmosphere and further raises high-altitude temperatures. In this vicious loop, the storm may spread over the entire planet, blanketing Mars in a pall more than fifteen miles deep. But as the darkening cloud blots out the Sun, the lower layers of the atmosphere cool and stabilize. Slowly the machinery of wind and sand dismantles itself until the cycle begins again.

In Sinai Planum—the Plain of Sinai—some twenty degrees south of the Martian equator, rust-colored dust is whipped into the Martian atmosphere by a complex interaction of local and hemispheric winds during the southern spring.

and rift valleys on Earth. The features are manifestations of a phenomenon known as plate tectonics—slow but inexorable movements of vast chunks of crustal material. Collisions and submersions of these chunks generate heat that powers chains of volcanoes near their edges. Further study was necessary, but some conjectured that Mars might be subject to a similar process.

A leading proponent of this theory was Bruce Murray, the affable geologist who later became director of JPL. Analyzing the images as they arrived, Murray maintained that the Tharsis activity was comparatively recent and that Mars, initially much like the Moon, was becoming Earth-like. Murray also expanded the discussion to include other aspects of the planet. He attributed its present-day atmosphere to geologic events, suggesting that the atmospheric gases had been belched out of volcanoes within the last one or two billion years. Unfortunately, he added, this point of view only made him more pessimistic about the possibility of life on the Red Planet: As bad as conditions are now, they could only have been worse in the distant past.

Other scientists looked at the same *Mariner 9* images and reached precisely the opposite conclusions. Carl Sagan, in particular, pointed to the Martian channels as evidence that perhaps as much as a billion years ago Mars had a much denser atmosphere and a climate that permitted the existence of liquid water. If so, he argued, life might well have evolved on Mars.

Even more intriguing, Sagan felt, were the closeup views of the polar caps *(pages 114-115)*. These thick deposits, which appeared to be at least partly made up of frozen water, had a striking wedding-cake appearance: Layers of ice were interleaved with dust. Apparently there had been many distinct depositions of ice and dust, each separated by relatively inactive periods. That this could happen in the present Martian climate, when the atmosphere is not thick enough to carry much dust to the poles and when the polar caps are actually being eroded by the relentless Martian winds, was a puzzle.

Sagan proposed a single explanation for both the channels and the polar laminations. Mars is known to experience periodic changes in its orbit and in the tilt of its rotational axis, brought about by the gravitational influences of the other planets. Sagan suggested that such changes could trigger climatic cycles, with arid, frigid conditions alternating over the span of about 100,000 years with a warm, humid environment. During the warm periods, the polar caps could melt; evaporating water would cause the atmospheric pressure to increase. Eventually, rain would fall and water would flow on the surface.

FOUR TRIES, THREE MISSES

To follow up on the findings of *Mariner 9*, the Soviet Union mounted a quadruple mission during the next launch period, in 1973. Only one of the Soviet probes was successful: *Mars 5* went into orbit around the planet and sent back photographs that supplemented the *Mariner 9* images but did little to improve upon them. Of the other three spacecraft, one missed Mars completely, another went into orbit and launched a landing vehicle that stopped transmitting during its descent, and the third flew past and ejected a lander that failed to hit the planet.

In the wake of this technological bust, planetary scientists were holding their breath on June 19, 1976, when the first of two U.S. Viking probes slipped into orbit around Mars. *Viking 1,* like the earlier Soviet spacecraft, was designed to place a module on the planet's surface, where it could directly sample the soil and atmosphere and take truly closeup pictures. The product of more than a decade of work by 10,000 people, the Viking probe was the most elaborate unmanned investigation of another world ever attempted. Just keeping track of its progress required the daily attention of a flight team that numbered 800 members.

Viking 1 spent a month reconnoitering the planet from orbit, determining that the landing site, selected after careful analysis of *Mariner 9* images, was too rough for a safe touchdown. A new landing point was chosen, and on July 20, the landing craft separated from the orbiter, 3,100 miles above the surface of Mars. From that moment onward, the lander was on its own. If anything went wrong, a cry for help would not reach mission control for nineteen minutes; a response would take the same time to make the 212-million-mile return trip.

Seven minutes after separation, the lander fired its retrorockets to begin a long, slow descent. Twenty-two minutes later, precisely according to schedule, the rockets cut off. Viking was three hours away from its target, a wide expanse of lava known as Chryse Planitia, the Plains of Gold. Not only did it appear to be relatively smooth, but Chryse was also judged geologically rich enough to be interesting. Many ancient channels had apparently dumped their water in this area.

The wait was excruciating. The utmost care had gone into selecting the landing site: Viking scientists had even tried to gauge its roughness with radar signals from the giant Arecibo radio telescope in Puerto Rico. But everyone was acutely aware that a surface that looked flat as a table from space could be crisscrossed by gullies large enough to swallow the lander whole. Among the Viking scientists, the unofficial odds on the success of the mission were no better than fifty-fifty.

The lander began to feel the effects of the Martian atmosphere at an altitude of 150 miles, while traveling almost horizontally at three miles per second. Brown University geologist Thomas (Tim) Mutch, head of the lander imaging team, later recalled the overwhelming silence of the moment as scientists listened to the mission controllers calling out each event. Within ten minutes, atmospheric friction had slowed the lander to a speed of 1,000 miles per hour; at 20,000 feet, it deployed a fifty-three-foot parachute to slow itself further. Then it extended its three legs and locked them into place. On the ground these legs would provide 8.7 inches of clearance. At JPL, people tried not to think about 10-inch rocks.

Just below 5,000 feet, the lander jettisoned the parachute, which had done as much as it could in the thin atmosphere. Almost immediately the craft fired its three terminal descent engines, carefully designed to disperse the exhaust plume and avoid contaminating the ground around the landing site. The lander's computer throttled the engines up and down to adjust its speed, using guidance from an on-board radar altimeter.

"Touchdown! We have touchdown!" The shout of Richard Bender, chief of the Viking lander engineers, rang through the JPL public-address system at 5:12 a.m. Despite the early hour, the exclamation triggered a cacophony of whoops, cheers, and applause, and then a party. Champagne corks popped. Scientists hugged and wept for joy. In mission control, controllers tore off their headphones and danced.

THE PLAINS OF GOLD

A few seconds after the landing, as insurance against some later equipment failure, the vehicle's computer automatically commanded the cameras to start taking their first pictures of Mars and relayed them to Earth through the orbiter overhead. At 5:47 a.m., the pictures began to arrive at JPL.

The dual cameras, housed in identical domes atop the lander, could provide a full 360-degree field of view as well as allow nearby objects to be photographed in stereo. The pictures were not delivered like conventional television images, however. The Viking cameras slowly built up their images, strip by vertical strip. Each individual scene took about twenty minutes to complete.

The first strip of the first picture showed little beyond the fact that the camera was working. Back at JPL the scientists and reporters watched the television monitors intently. The second strip showed a rock. It was ordinary, unspectacular, only a few inches across. The picture built up into a crisp black-and-white image of the pad of the lander's third leg, resting on pebbly Martian soil. Tim Mutch later remembered thinking that the cameras seemed to be working better than they had on Earth.

On July 20, 1976, members of the Viking biology team were transfixed by a television monitor at the Jet Propulsion Laboratory *(above)* as *Viking 1* set down on Mars. Moments later the first closeup view of a Martian landscape was complete *(below):* a desert of sand and rock at Chryse Planitia.

Everyone waited with rapt attention: The second picture was to be a full panorama. Mutch had prepared for this moment by traveling to Antarctica, the place on Earth where conditions are believed to be most similar to those on Mars. Trying to get a preview of what Viking would encounter, he collected and studied a wide variety of geologic specimens on the frozen continent. His favorites were "ventifacts," rocks carved and polished by the wind until they looked like the abstract creations of a modern sculptor. If such things turned up on Mars, Mutch warned, they should not be taken as evidence of intelligent life.

Slowly the image built up: rocks, boulders, more rocks, windblown silt, dunes . . . a bright sky and brilliant sunlight . . . more rocks, pitted and eroded. To the geologists, the Plains of Gold looked less like Antarctica than like a terrestrial desert—a barren, sunlit wasteland. To Mutch it was lovely: "It's just a beautiful collection of boulders," he declared, "a geologist's delight! This is just incredible good luck."

The good fortune had another aspect. Only twenty-five feet away from the vehicle was a boulder ten feet across and three feet high, later nicknamed

Big Joe. With a difference of only a few yards in the descent, the *Viking 1* lander might have ended its mission in complete disaster.

Hundreds of images followed in succeeding weeks. For the scientists, it was the beginning of a deliriously happy period of discovery, when every observation brought new details to light and every detail spoke volumes. The endless rocks, for example, testified to the rich and complex history of Mars: They might have been hurled there by a volcano, washed down from the mountains by an ancient torrent of water, or ejected from an impact crater. As team geologist Alan Binder explained it, "I'd expected to see maybe half a dozen rocks, all pretty similar. And there was this forest of thousands of rocks. We'll never analyze them all, but I see at least thirty kinds."

Then there was the Martian sky, rendered an unearthly salmon color by fine dust suspended in the atmosphere. The lander's miniature chemical laboratory demonstrated that Mars came by its rust-red color honestly: The soil was largely composed of iron oxides. Silicon was also abundant. Overall, the rocks seemed to resemble terrestrial basalt—just what one might expect from a lava plain.

The climate, as predicted, was forbidding. The weather report from the *Viking 1* station, as it was now called, indicated that the temperature on the sunny-looking rockscape ranged from a low of minus 122 degrees Fahrenheit

In three panoramic images, the cameras of *Viking 2* reveal a plain of volcanic rock, boulders, and sand at Utopia Planitia in Mars's northern hemisphere. The lander, which arrived on September 3, 1976, found the soil to be rich in sulfur, carbon, iron, and iron hydroxides, source of the well-known reddish cast.

just after dawn to a balmy maximum of minus 22 degrees around noon. The wind was less inclement than the cold: In those early days it was never more than fourteen miles per hour. Garry Hunt of the University of London gave a tongue-in-cheek (but completely accurate) forecast: "fine and sunny; very cold; winds light and variable; further outlook similar."

As for the atmosphere, the lander's sensors, working throughout the descent, confirmed that it was 95 percent carbon dioxide. *Viking 1* also detected nitrogen for the first time; this was good news for those who still hoped to find Martian life, since nitrogen is one of the four elements essential to life as it exists on Earth. The other elements—carbon, oxygen, and hydrogen—had all been detected previously.

Geophysicist Michael McElroy of Harvard University argued that this present-day nitrogen was solid evidence for a much heavier atmosphere perhaps three billion years in Mars's past. Over the eons, he said, massive amounts of the nitrogen escaped into space. So perhaps it was true that Mars once had an Earth-like atmosphere—warmer, heavy enough to let rain fall, and dense enough to let life arise in some unknown sea.

But none of this data was solid evidence for the existence of Martian life now. That would have to come from the star attraction on the Viking lander, a miniaturized chemical apparatus designed to test the soil for the presence of living microbes. The $60-million biology laboratory was the prime reason for public interest in and support for the Viking project. However, the results, far from settling the question, were both bizarre and ambiguous.

In one of the three biology experiments, for example, a tiny sample of soil was soaked with a nutrient solution, dubbed "chicken soup" by the scientists. The soil suddenly started producing billows of oxygen, just as though tiny Martian organisms had been provoked into a frenzy of activity. But the initial oxygen output was fifteen times more than even optimists had expected, and after two days it leveled off and then began to drop—instead of continuing to increase, as microbe-produced oxygen would.

MARTIAN MIMICRY

Adding to the mystery were the results of tests done by a piece of equipment known as the gas chromatograph/mass spectrometer, which analyzed the molecular content of the soil directly. In particular, it looked for complex carbon molecules that might have been left behind by generations of dead microbes. Such molecules are abundant in the soil of Earth, even in desolate Antarctica. But the Viking instrument found nothing of the kind. At best, it seemed that some exotic chemicals in the Martian soil might be mimicking the activities of life.

Scientists now looked to the *Viking 2* lander to resolve these uncertainties. The craft touched down on September 3 on the broad, flat Utopia Planitia, 4,600 miles northeast of Chryse. Utopia's topography was a surprise. Instead of the gently rolling sand dunes and dust indicated by pictures taken from orbit, it was a rocky plain much like that seen by *Viking 1.* The results the second lander obtained from its biology experiments were as equivocal as those of its twin. "You might say that we're in the seventh inning and the score is Biology 10, Exotic Chemistry 10," said a baffled Gerald Soffen, chief scientist of the Viking team.

As days and then months passed, it became clear that the Viking landers would not yield conclusive proof one way or the other on the question of Martian life. But the landers were immensely valuable on other fronts. For one thing, they lasted far longer than anyone had dared hope. The *Viking 1* station, for example, was still operating in October 1980, when Tim Mutch died in a

A thin layer of frost brightens the ground at Utopia Planitia. Although the atmosphere of Mars is too cold in winter—down to minus ninety-two degrees Fahrenheit—to hold much moisture, scientists conclude that water vapor on dust blown from the warmer south could settle with freezing carbon dioxide into a frosty coating on the sand.

mountain-climbing accident. (The landing site was officially named the Thomas A. Mutch Memorial Station, and the geologist's colleagues hoped that one day astronauts would visit that spot on the Plains of Gold and leave a plaque with a dedication.)

During their service, the Vikings sent back streams of weather reports. They imaged the changing seasons—there is frost on the ground in winter—and even listened for Marsquakes, which would indicate that geologic activity continues beneath the planet's surface. By the time the *Viking 1* lander stopped transmitting in November 1982, it had produced enough data to keep planetary scientists busy for years to come.

The final transmission from *Viking 1* was the last direct information scientists would have of Mars for more than a decade. The Soviets, after their repeated failures at Mars, turned their attention to a much more successful series of missions to Venus. The United States, limited by budgetary cutbacks and concentrating on its space shuttle program, essentially dropped out of planetary exploration.

RED PLANET REDUX

In July 1988, Mars returned to the limelight when the Soviet Union sent *Phobos 1* and *Phobos 2* to investigate the Red Planet and the larger of its two satellites. Neither craft completed the mission, and *Phobos 2* returned only a few images of the target moon before losing contact with Earth. But the two probes marked the beginning of a planned decade-long exploration program, including a sophisticated mission in 1994 that would feature the release into the Martian atmosphere of a set of French-built balloons. At night, the balloons would cool and sink, lowering their instrument packages to the surface. Warmed by sunlight, they would rise again to take panoramic pictures of the Martian landscape. The program is intended to culminate with a sample-return mission, in which robot vehicles will rove the surface, picking up rocks to bring home to Earth. NASA, too, will return to Mars in the 1990s. Its Mars Observer spacecraft, scheduled for a 1992 launch, is designed to orbit the planet, mapping its surface, investigating its chemical composition, and looking for evidence of water.

During the late 1980s, the idea of a joint Soviet-U.S. mission to Mars was touted by scientists and politicians alike. Such a venture was even discussed at summit conferences between leaders of the two nations, ordinarily competitors in space. Carl Sagan, ever the optimist, joined many others to push the idea of a joint manned mission early in the next century.

A manned voyage would cost tens of billions of dollars and require humans to endure the cramped confines of space vehicles and planetary habitats for up to three years, far beyond the record-setting one-year stints of Soviet cosmonauts in Earth's orbit. But proponents of such an expedition maintain that it is the only way to resolve many issues raised by the Mariner and Viking missions. Mars has a surface area equal to the entire land area of the Earth, and only two points have been closely examined. Humans on Mars could roam far and wide, gathering rocks and soil, taking core samples, setting up seismic experiments, and most important, following up hunches.

Their scientific agenda would be shaped by broad lines of inquiry, with a sharp focus on the planet's past. The ancient atmosphere is one target. Explorers could drill holes at carefully selected sites, including the polar caps, and extract samples whose chemical composition would provide valuable clues in the investigation of Martian climatic cycles. Similar studies could detect the extent of permafrost below the surface; knowledge of the current distribution of water could give insight into the early epoch when free-flowing water carved out the now-dry valleys.

Geology would be another important area of scrutiny. Expeditions around the Tharsis Bulge and the Valles Marineris rift could help answer questions about Martian tectonics and other geologic activity. Widely spaced seismographs could monitor movement in the crust and reveal more about the core of the Red Planet.

Certainly the longstanding question of Martian life would receive scrupulous attention. A paleontologist walking across the dusty plains might find fossils in the rocks, even if just fossilized bacteria. Explorers might actually come across tenacious oases of life around scattered hot springs, where the planet's internal heat melts the permafrost.

Whatever the outcome of future missions, Mars is likely to keep its hold on the human imagination—but in a new way. No longer is it a mirror reflecting earthly fears and fantasies. Its very kinship with Earth, like that of Mercury and Venus, suggests how subtle are the forces that have shaped the Third Planet's life-giving environment. In their various ways, all help define the planet we call home.

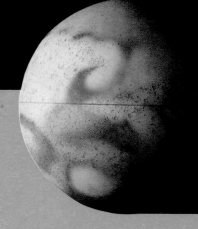

A WORLD DIVIDED

When the Mariner and Viking missions surveyed the distant environs of Mars in the 1970s, they revealed an austere planet of rusty, rugged beauty. Though just over half Earth's size, Mars is more than a match in the variety and scale of its landforms. Canyon, crater, volcano, ice field, and plain combine in sometimes fantastic proportions to make Mars a world of great visual drama.

Mars and Earth are sibling planets, born at the same time and of similar materials. Many of the geologic processes active on Earth have apparently shaped the Martian landscape as well. Volcanism, erosion, sedimentation, and tectonic forces that shift the planetary crust have all helped sculpt the Red Planet. In addition, meteorites have pitted broad expanses of its surface.

These processes have worked selectively, producing two hemispheres with distinctly different terrains. To the south lie heavily cratered highlands. To the north spread smoother plains, nearly two miles lower in mean elevation. The south is predominantly an ancient landscape, little changed since the planet was assailed by meteorites early in its history. The northern lands, by contrast, have been greatly transformed through the eons. Some areas are covered with lava flows or spotted with small volcanic cones. In others, wind- or water-borne sediments have created dunes and drifts. Most spectacular of all are features that straddle the demarcation line between the two hemispheres: immense canyonlands and giant volcanoes, testifying to the power of the forces that have molded the planet's face.

MONSTER VOLCANOES

More than half of Mars's frigid surface is a stony reminder of the fiercely hot magma deep below. Lava fields and chunks of debris from ancient eruptions cover many of the northern lowlands, and nowhere are the effects of the planet's inner fire more evident than in the volcanic peaks of the Tharsis Bulge. A mammoth uplift near the equator, Tharsis rises four miles above the planet's average elevation. At its crown are volcanoes that dwarf any in the Solar System. The mightiest, Olympus Mons, has more than fifty times the bulk of Hawaii's Mauna Loa, Earth's largest. Like the Hawaiian peak, the Martian titans are shield volcanoes, with conical shapes and gentle slopes of about six degrees. They also show similar lava types and eruption patterns and are topped with calderas—large summit craters caused by the collapse of the central cone when lava flows leave huge voids below the surface. Unlike terrestrial volcanoes, however, the Martian cones can apparently grow indefinitely. On Earth, eruptions are regulated by the movement of plates—huge blocks of the planet's crust—that eventually isolate the volcano from the molten rock below. There is no evidence of plate movement on Mars, leading scientists to conclude that a volcano there can continuously tap the same column of rising magma and grow to enormous size.

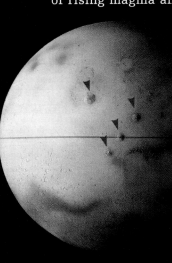

In the image at right, taken by a Viking orbiter in 1976, huge volcanoes sprawl across the Tharsis region, indicated by the arrows on the globe at left. Aligned north to south astride the Tharsis ridge *(center right, top to bottom)* are Ascraeus Mons, Pavonis Mons, and Arsia Mons, all towering seventeen miles above the average surface level. To the left of the trio is Olympus Mons, taller and far broader—the monarch of Martian volcanoes.

The collapsed roof of a subsurface magma chamber creates a fifty-mile-wide summit caldera on Olympus Mons. The vast crater grew in stages as molten rock poured out over a span of a billion years, emptying the cavern below.

A vast flood narrowed to form this outflow channel, curving around a meteorite crater.

A short stretch of a 500-mile-long runoff channel shows a webwork of steep-sided tributaries, probably carved by water that seeped from below the surface.

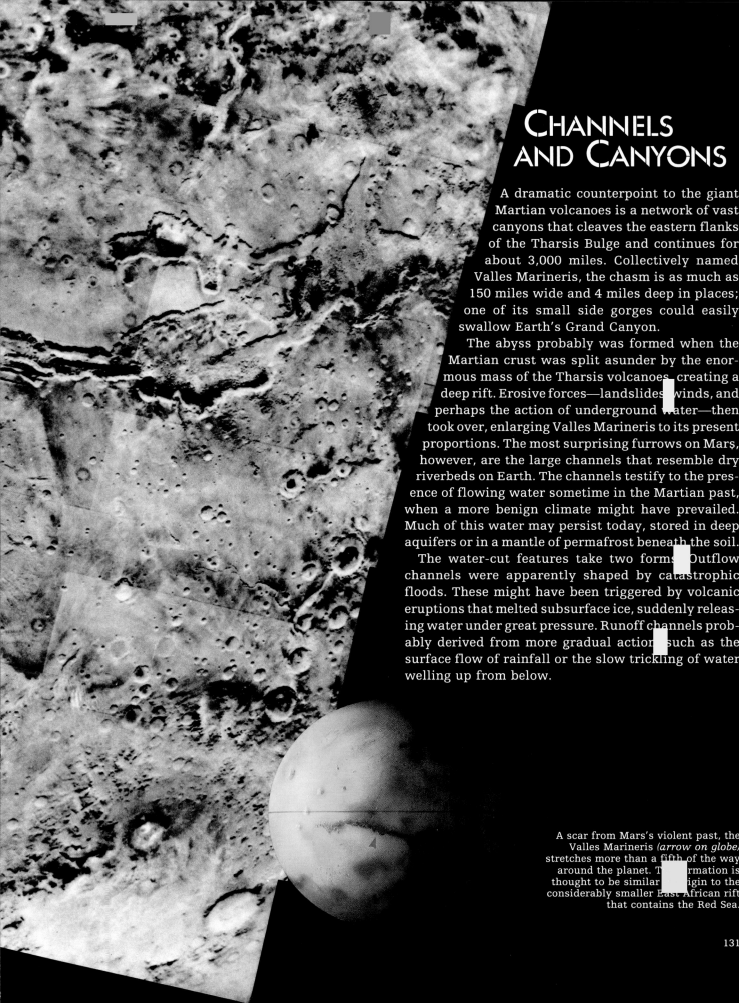

CHANNELS AND CANYONS

A dramatic counterpoint to the giant Martian volcanoes is a network of vast canyons that cleaves the eastern flanks of the Tharsis Bulge and continues for about 3,000 miles. Collectively named Valles Marineris, the chasm is as much as 150 miles wide and 4 miles deep in places; one of its small side gorges could easily swallow Earth's Grand Canyon.

The abyss probably was formed when the Martian crust was split asunder by the enormous mass of the Tharsis volcanoes, creating a deep rift. Erosive forces—landslides, winds, and perhaps the action of underground water—then took over, enlarging Valles Marineris to its present proportions. The most surprising furrows on Mars, however, are the large channels that resemble dry riverbeds on Earth. The channels testify to the presence of flowing water sometime in the Martian past, when a more benign climate might have prevailed. Much of this water may persist today, stored in deep aquifers or in a mantle of permafrost beneath the soil.

The water-cut features take two forms. Outflow channels were apparently shaped by catastrophic floods. These might have been triggered by volcanic eruptions that melted subsurface ice, suddenly releasing water under great pressure. Runoff channels probably derived from more gradual action, such as the surface flow of rainfall or the slow trickling of water welling up from below.

A scar from Mars's violent past, the Valles Marineris *(arrow on globe)* stretches more than a fifth of the way around the planet. The formation is thought to be similar in origin to the considerably smaller East African rift that contains the Red Sea.

SOUTHERN CRATERING

The southern ice cap, one of the youngest geologic features on Mars, sparkles against a dull terrain pocked with ancient craters. Built up in layers and laced with a swirling complex of terraced valleys that are more than half a mile deep, the polar formation looks from space like a huge, misshapen pinwheel.

Scientists surmise that the layers, which are present at the planet's north pole as well, result from the seasonal deposition and melting of dust-laden carbon dioxide ice; the dust is left behind when the volatile carbon dioxide evaporates. Individual strata, some of them more than 300 feet thick, are thought to represent recent climatic periods. The origin of the spiral valleys is a puzzle, however; wind erosion and varying patterns of melting are believed to be important factors.

The cratered terrain around the south pole is thought to be a remnant of the final era of planetary formation, when the rate of collision with meteorites was high. Some of the impact craters are unlike any others in the Solar System. Known as splosh craters, they are surrounded by ejected material that seems to have had a mudlike consistency, as if water or ice lay near the surface at the point of impact.

A swirling white cap *(far right)* of frozen carbon dioxide—possibly admixed with water ice—lies slightly askew of the southern pole *(bottom of globe).* More than 200 miles across in this summer picture, the polar cap expands to more than twelve times that size in winter.

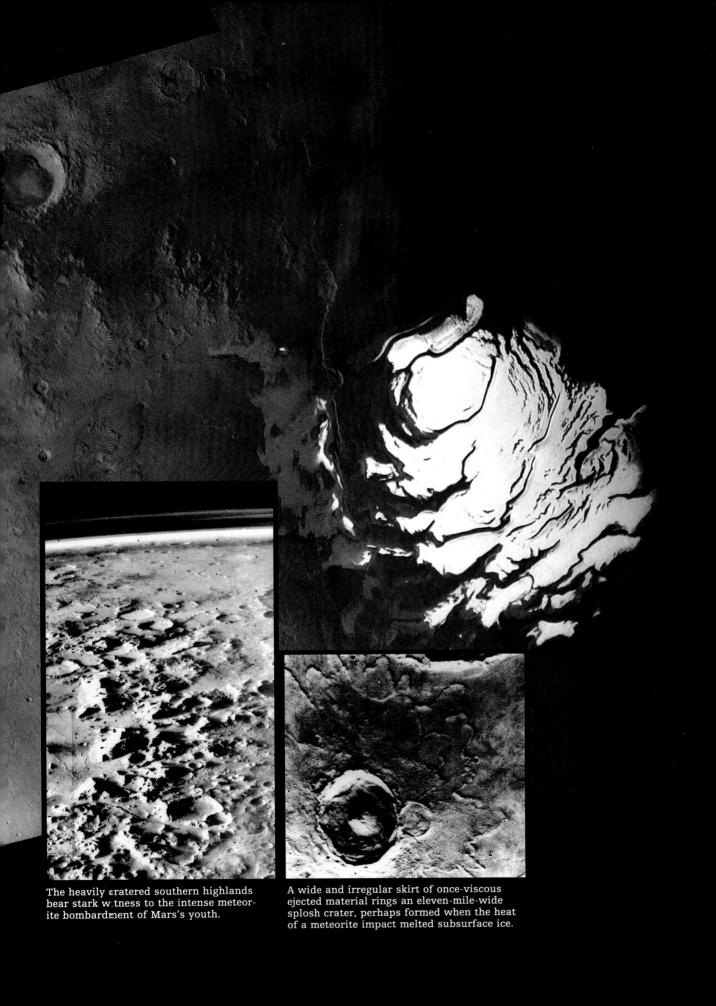

The heavily cratered southern highlands bear stark witness to the intense meteorite bombardment of Mars's youth.

A wide and irregular skirt of once-viscous ejected material rings an eleven-mile-wide splosh crater, perhaps formed when the heat of a meteorite impact melted subsurface ice.

APPENDIX

DENIZENS OF THE SOLAR SYSTEM

Pluto

Mercury

Venus

Saturn · Mars · Neptune · Jupiter · Uranus · Earth

Formed out of material circling the Sun, the planets ultimately settled into orbits slightly angled to the plane of the ecliptic, the apparent path of the Sun across the sky. As seen above, the orbits of Mercury and Pluto, which are the most elliptical in the Solar System, are also the most inclined.

Although all of them are ruled by the same physical laws, there is considerable difference among the Sun's planets. For instance, they orbit at varying distances from the Sun, in paths shaped by the balance between solar gravity and their own motion. The way they are tipped with respect to their orbits largely defines their seasons, and their individual rates of spin set the length of each planetary day.

Scientists do not yet know the underlying causes of some planetary phenomena—why Uranus wallows around its orbit on its side, for example, or why Pluto's orbit differs so radically from those of its fellows—but much has been learned about these bodies over the years. On these two pages are diagrams that depict the planets' sizes, orientations, and orbits and a table *(right)* that highlights the salient traits of the denizens of the Solar System.

Pluto

Neptune

Saturn

Jupiter

Uranus

From a vantage above the north pole of the Sun, the orbits of most of the planets appear to be nearly circular. Pluto's orbit, in contrast, is so elliptical that the planet swings inside the orbit of Neptune for about 20 of every 248 years.

Mars

Earth

Mercury

Venus

Asteroid Belt

Planetary Data	Mercury	Venus	Earth	Mars	Jupiter	Saturn	Uranus	Neptune	Pluto
Equatorial Diameter (Miles)	3,031	7,521	7,926	4,217	88,730	74,600	31,600	30,700	1,420
Mass (Trillion Trillion Pounds)	0.729	10.738	13.177	1.416	4,187.0	1,253.8	190.95	226.0	0.026
Mean Density (Earth = 1)	0.98	0.95	1.0	0.71	0.24	0.125	0.216	0.286	0.36
Gravity (Earth = 1)	0.39	0.88	1.0	0.38	2.34	0.93	0.79	1.09·	0.0637
Period of Rotation (Hours)	1,407.6	5,832.2	23.9	24.6	9.8	10.2	17.2	16.1	6.4
Escape Velocity (Miles per Hour)	9,619	23,042	25,055	11,185	133,104	79,639	47,470	52,558	2,640
Major Atmospheric Gas	Oxygen	Carbon Dioxide	Nitrogen	Carbon Dioxide	Hydrogen	Hydrogen	Hydrogen	Hydrogen	Methane
Inclination of Equator (Degrees)	0.0	2.6	23.5	25.2	3.1	26.7	82.1	29.0	62.0
Known Moons	0	0	1	2	16	17	15	8	1
Eccentricity of Orbit	0.206	0.007	0.017	0.093	0.048	0.056	0.047	0.009	0.246
Mean Orbital Velocity (Miles per Hour)	107,132	78,364	66,641	53,980	29,216	21,565	15,234	12,147	10,604
Minimum Distance from Sun (Millions of Miles)	28.6	66.8	91.4	128.4	460.3	837.6	1,699.0	2,771.0	2,756.0
Maximum Distance from Sun (Millions of Miles)	43.4	67.7	94.5	154.9	507.2	936.2	1,868.0	2,819.0	4,555.0
Mean Distance from Sun (Millions of Miles)	36.0	67.2	93.0	141.6	483.4	886.7	1,784.0	2,794.4	3,656.0
Period of Revolution (Earth Years)	0.24	0.62	1	1.88	11.86	29.46	84.01	164.79	247.70
Inclination of Orbit to Plane of Ecliptic (Degrees)	7.00	3.39	—	1.85	1.31	2.49	0.77	1.77	17.15

All of the planets except Mercury rotate around an axis tilted with respect to the plane of their orbit (above). Mercury's axis is not inclined, while Uranus and Pluto display the most extreme inclinations, spinning almost on their sides.

Diagramed below, the planets' average relative distances from the Sun stretch more than three billion miles from the asteroid belt to the distant track of Pluto.

GLOSSARY

Angular diameter: an object's width on the sky, measured in units of arc distance. The Moon's angular diameter is just over half a degree; Mars's is 3.7 seconds of arc at the planet's nearest approach to Earth, 17.9 seconds of arc at conjunction.

Aphelion: the point in the orbit of a planet or a comet where it is farthest from the Sun.

Apparent day: the period between one high noon and the next, as observed from a particular point on a planet's surface; also called solar day, or sol. *See* Sidereal day.

Arc distance: units used in measuring the apparent position and size of celestial bodies. Arc distance is expressed in degrees, minutes (sixty per degree), and seconds (sixty per minute).

Atmosphere: a gaseous shell surrounding a planet or other body; also, a unit of pressure equal to the amount of atmospheric pressure at sea level on Earth, about 14.7 pounds per square inch.

Axis: the imaginary line, drawn through the poles of a celestial body, around which the body rotates; also, one of two perpendicular lines (the major axis and the minor axis) passing through the center of an ellipse.

Basalt: a dark, close-grained igneous rock, formed by the hardening of lava. Most volcanic rocks are basalts.

Bow shock: in planetary science, the boundary region where the solar wind is first deflected by a planet's magnetic field.

Caldera: a crater formed by the collapse or subsidence of the central part of a volcano.

Celestial mechanics: the study of the motions and interactions of astronomical objects.

Compression wave: a seismic wave that expands and then contracts the medium through which it passes.

Conjunction: an apparent meeting or close approach of two celestial bodies as viewed from Earth. In planetary astronomy, if only one planet is mentioned, the second body in the conjunction is the Sun. Inferior conjunction occurs when Venus or Mercury passes between the Sun and Earth; superior conjunction occurs when a planet is directly behind the Sun.

Convection: the transfer of heat in a fluid or gas by the movement of currents from hotter to cooler regions.

Core: the central component of a celestial body. In planets, it is usually composed of dense, hot material, frequently solid.

Corona: the outer layer of the Sun's atmosphere, composed of diffused, ionized gas.

Coronae: moundlike structures on Venus, possibly caused by crustal expansion, averaging one-half-a-mile high and 100 to 350 miles in diameter.

Crater: a bowl-shaped depression on a planet's surface, formed by the impact of a meteorite or comet.

Crust: the solid surface layer of a planet.

Deuterium: an isotope of hydrogen, sometimes referred to as heavy hydrogen. A hydrogen nucleus contains a single proton; a deuterium nucleus contains one proton and one neutron.

Doppler shift: a change in the wavelength and frequency of sound or electromagnetic radiation, caused by the motion of the emitter, the observer, or both.

Eccentricity: the amount of separation between the two foci of an ellipse and, hence, the degree to which the ellipse deviates from a circular shape. Mercury's elongated orbit, for example, is more eccentric than Earth's.

Eclipse: the obscuration of light from a celestial body as it passes through the shadow of another body.

Ejecta: the matter that is thrown from a depression on the surface of a planet, either through a meteorite or comet collision or through a volcanic eruption or other tectonic process.

Electromagnetic spectrum: the array, in order of frequency or wavelength, of electromagnetic radiation, from low-frequency, long-wavelength radio waves to high-frequency, short-wavelength gamma rays.

Ellipse: a closed, symmetrical curve drawn so that the sum of the distances from any point on the curve to each of two fixed points (called foci) is constant. A circle is an ellipse with its two foci superimposed at the center. The orbits of all Solar System planets are ellipses.

Elongation: the separation of two celestial bodies as seen from a third, expressed in units of arc distance. The elongation of a planet refers to its distance from the Sun, as measured from Earth.

Fault: a surface fracture where rock on one side has moved in relation to the rock on the other side. Fault movement may be vertical, horizontal, or oblique.

Flux ropes: lines of magnetic force found in the ionosphere of Venus.

Gas chromatograph: a device for the analysis of chemical compounds or mixtures of compounds.

Gravity assist: the use of a celestial body's gravitational force to alter the trajectory of a spacecraft. Gravity-assist maneuvers allow speed and direction changes that are otherwise impossible for missions with limited fuel.

Greenhouse effect: a phenomenon in which heat is trapped near the surface of a planet by atmospheric gases and clouds. Short-wavelength solar radiation readily penetrates the atmosphere, but long-wavelength infrared radiation from the heated surface is absorbed, thereby causing gradual warming.

Hadley cell: a cycle in a planetary atmosphere in which convection pushes rising gas from warm areas of the atmosphere to cooler ones, where the gases sink, circulate back to the warm areas, and are again pushed upward.

Hydrometer: an instrument that measures the specific gravity of a liquid, usually to help determine its composition.

Inferior planet: a planet whose orbit falls between the Sun and Earth's orbit, specifically Mercury or Venus.

Infrared: a band of electromagnetic radiation with a lower frequency and a longer wavelength than visible red light.

Ion: an atom that has lost or gained one or more electrons, giving it a positive or negative electrical charge. A neutral atom has an equal number of electrons and protons, giving it a zero net electrical charge. A positive ion of an element has fewer electrons than the neutral atom; a negative ion has more.

Ionopause: the boundary area at the upper limit of an ionosphere.

Ionosphere: an atmospheric layer dominated by charged, or ionized, atoms.

Kepler's laws: mathematical principles describing the motion of planets around the Sun, developed by seventeenth-century German astronomer Johannes Kepler. The first law states that every planetary orbit is an ellipse, with the Sun at one focus. Kepler's second law states that a line connecting a planet to the Sun sweeps out equal areas of space in equal intervals of time—that is, the planet moves faster when it is nearer to the Sun. The third law expresses the direct relationship between a planet's average distance from the Sun and the period of its orbit.

Lava: molten or semimolten rock issuing from vents or cracks on the surface of a planet.

Magma: molten rock formed beneath the surface of a planet.

Magnetometer: a device for measuring the strength and direction of a magnetic field.

Mantle: the layer of a planet between the outer core and the crust.

Mass spectrometer: a device used to determine a substance's chemical composition by measuring the varied masses of its components.

Meander: a loop-shaped curve formed by water action, as in a river channel.

Meridian: an imaginary north-south line in the sky directly over any point on a planet's surface.

Meteoroid: a small metallic or rocky body found in space. A meteoroid entering a planet's atmosphere is called a meteor. Meteors often burn up in the atmosphere; those that reach the surface are termed meteorites.

Molecule: the smallest component of a chemical that retains the chemical's properties. A molecule may consist of a single atom or, more commonly, two or more atoms bonded together.

Moon: one of a planet's natural satellites, generally no smaller than ten miles in diameter. There are more than fifty known moons in the Solar System.

Opposition: the alignment of two celestial bodies on opposite sides of the sky as seen from Earth. The opposition of a superior planet occurs when Earth passes between it and the Sun; a perfect opposition occurs when a planet is also at its closest approach to Earth, at its best point for observation.

Orbit: the path of an object revolving around another object.

Perihelion the point in the orbit of a planet or a comet where it is closest to the Sun.

Permafrost: ground that is permanently frozen unless artificially heated.

Phase: one of the recurring appearances of a celestial body viewed from Earth. As an object such as the Moon or Venus moves along its orbit, the amount of its surface visible from Earth increases (waxes) and decreases (wanes) at a regular, periodic rate.

Planet: a large, nonstellar body that orbits a star and shines only with reflected light.

Planetesimal: in theory, a small orbiting body that actively accretes mass from random collisions and will eventually become a full-scale planet.

Plasma: a gaslike association of ionized particles that responds collectively to electric and magnetic fields. Because plasma particles do not interact the way particles of ordinary gas do, plasma is considered a fourth state of matter, along with solid, liquid, and gas.

Polar caps: the distinctive surface areas near the poles of Earth and Mars. On Mars, the polar caps are mainly frozen carbon dioxide, perhaps including some water ice. The Martian polar caps shrink and expand with the seasons.

Polar hood: a dense cloud cover sometimes seen over the polar caps on Mars.

Probe: an automated, unmanned spacecraft used to collect data and transmit it to Earth.

Proto-Sun: in theory, the gaseous matter at the center of the Solar nebula, held together by its own gravitational attraction, that shrank and compressed to become the Sun about 4.5 billion years ago.

Pyrite: a metallic yellow mineral compound of iron and sulfur.

Radar: a method of identifying the location or speed of a distant object by bouncing radio waves off its surface and measuring the interval before they return; also, an instrument used for this purpose. The term is an acronym for "radio detection and ranging."

Radiation: energy in the form of electromagnetic waves or subatomic particles.

Radio: the least energetic form of electromagnetic radiation, having the lowest frequency and longest wavelength.

Radio altimeter: an instrument that determines altitude by bouncing radio waves off the surface below and measuring the interval before they return.

Radio astronomy: the observation and study of radio waves produced by astronomical objects.

Radiometer: a device that measures the intensity of radiation.

Radio telescope: an instrument for studying astronomical objects at radio wavelengths.

Retrograde motion: real or apparent motion against the prevailing direction of movement. The apparent retrograde motion of Mars as seen from Earth is an illusion, produced by the combined effect of the orbital motion of each planet. Venus's retrograde rotation is real.

Retrorocket: a small rocket used to decelerate a larger rocket or a satellite.

Revolution: the orbiting of one celestial body around another; a single cycle of such a movement.

Rotation: the turning of a celestial body about its axis.

Satellite: any body, natural or artificial, in orbit around a planet; used most often to describe moons and spacecraft.

Scarp: a steep, clifflike ridge, formed by faulting or by erosion of the softer rock layers of tilted formations.

Sediment: mineral particles deposited by wind or water action. Layers of sediment may consolidate to form rock.

Seismic wave: a vibration through solid rock, triggered by an earthquake or artificial means, that extends in all directions from the point of the initial disturbance.

Seismograph: a device for recording the strength and frequency of seismic waves.

Sidereal day: the period of time it takes a planet to rotate once on its axis, measured from one appearance of a chosen fixed star on the meridian to the next such appearance. *See* Apparent day.

Silicate: a mineral based on silicon and oxygen, usually containing one or more other elements. Silicates are the primary components of most rocks.

Sol: *See* Apparent day.

Solar nebula: the disk of spinning gas and dust from which the Sun and planets formed.

Solar wind: a current of charged particles that streams outward from the Sun.

Spectroscopy: the study of electromagnetic spectra, including the position and intensity of emission and absorption lines, to learn about the physical processes and chemical compositions that create them.

Spectrum: the array of colors or wavelengths obtained by dispersing electromagnetic radiation, as when light is passed through a prism. A spectrum is often banded with absorption or emission lines, which can be interpreted to reveal the chemistry and motion of the radiation source.

Sulfuric acid: a corrosive compound of sulfur, hydrogen, and oxygen. A major component of the Venusian atmosphere, it occurs in a dilute form on Earth as acid rain.

Superior planet: a planet whose orbit is farther from the Sun than Earth's.

Synchronous rotation: a phenomenon in which a moon spins on its axis exactly once per orbit, thus keeping the same face toward its planet at all times; seen in most moons in the Solar System.

Tectonics: the study of a planet's crust, including its structure and processes.

Telemetry: the data, usually measurements, transmitted from a remote sensor to a recording receiver.

Tidal slowing: the deceleration of rotation in a moon or planet caused by the gravitational pull of other bodies, primarily the Sun.

Tide: a change in the shape of one body resulting from the gravitational pull of another.

Transit: the passage of a small celestial body across the visible face of another, larger body; also, the movement of any celestial body across a meridian.

Ultraviolet: a band of electromagnetic radiation with a higher frequency and shorter wavelength than visible blue light. Most ultraviolet radiation is absorbed by Earth's atmosphere, so ultraviolet astronomy is normally performed in space.

Weird terrain: the name given to an area on Mercury that is characterized by unusual hill and valley formations, thought to have formed as a result of a single giant impact on the other side of the planet.

BIBLIOGRAPHY

Books

Abbott, David, *The Biographical Dictionary of Scientists: Astronomers.* New York: Peter Bedrick Books, 1984.

Abell, George O., David Morrison, and Sidney C. Wolff, *Exploration of the Universe.* Philadelphia: Saunders College Publishing, 1987.

Audouze, Jean, and Guy Israël, eds., *The Cambridge Atlas of Astronomy.* Cambridge: Cambridge University Press, 1985.

Barbato, James P., and Elizabeth A. Ayer, *Atmospheres: A View of the Gaseous Envelopes Surrounding Members of Our Solar System.* New York: Pergamon Press, 1981.

Baugher, Joseph F., *The Space-Age Solar System.* New York: John Wiley & Sons, 1988.

Biemann, Hans-Peter, *The Vikings of '76.* Cambridge, Mass.: Hans-Peter Biemann, 1977.

Burgess, Eric, *Venus: An Errant Twin.* New York: Columbia University Press, 1985.

Carr, Michael H.:
"The Geology of Mars." In *The Solar System and Its Strange Objects,* ed. by Brian J. Skinner. Los Altos, Calif.: William Kaufmann, 1981.
The Surface of Mars. New Haven, Conn.: Yale University Press, 1981.

Carr, Michael H., and Nancy Evans, *Images of Mars: The Viking Extended Mission.* Washington, D.C.: NASA, 1980.

Carr, Michael H., et al., *The Geology of the Terrestrial Planets.* Washington, D.C.: NASA, 1984.

Chapman, C., and F. Vilas, eds., *Mercury.* Tucson: University of Arizona Press, 1988.

Chapman, Clark R., *Planets of Rock and Ice.* New York: Charles Scribner's Sons, 1982.

Davies, Merton E., et al., *Atlas of Mercury.* Washington, D.C.: NASA, 1978.

Debus, Allen G., ed., *World Who's Who in Science.* Wilmette, Ill.: Marquis Who's Who, 1968.

De la Cotardière, Philippe, *Larousse Astronomy.* New York: Facts on File Publications, 1987.

Dunne, James A., and Eric Burgess, *The Voyage of Mariner 10: Mission to Venus and Mercury.* Washington, D.C.: NASA, 1978.

Fimmel, Richard O., Lawrence Colin, and Eric Burgess, *Pioneer Venus.* Washington, D.C.: NASA, 1983.

Firsoff, V. A., *The Solar Planets.* New York: Crane, Russak, 1977.

French, Bevan M., *Mars: The Viking Discoveries,* Washington, D.C.: NASA, 1977.

Furniss, Tim, *Jane's Spaceflight Directory: 1986.* London: Jane's Information Group, 1983.

Glasstone, Samuel, *The Book of Mars.* Washington, D.C.: NASA, 1968.

Greeley, Ronald, *Planetary Landscapes.* Boston: Allen & Unwin, 1987.

Hartmann, William K., and Odell Raper, *The New Mars: The Discoveries of Mariner 9.* Washington, D.C.: NASA, 1974.

Herrmann, Dieter B., *The History of Astronomy from Herschel to Hertzsprung.* New York: Cambridge University Press, 1984.

Hunt, Garry E., and Patrick Moore, *The Planet Venus.* London: Faber and Faber, 1982.

Joëls, Kerry Mark, *The Mars One Crew Manual.* New York: Ballantine Books, 1985.

Kane, John T., "Asaph Hall and the Moons of Mars." In *Wanderers in the Sky,* ed. by Thornton Page and Lou Williams Page. New York: Macmillan, 1965.

Kaufmann, William J., III, *Universe.* New York: W. H. Freeman, 1987.

Kopal, Zdenek, *Realm of the Terrestrial Planets.* London: Institute of Physics, 1979.

Krasnopol'sky, V. A., and V. A. Parshev, "Photochemistry of the Venus Atmosphere." In *Venus,* ed. by D. M. Hunten et al. Tucson: University of Arizona Press, 1983.

Lewis, John S., and Ronald G. Prinn, *Planets and Their Atmospheres: Origin and Evolution.* Orlando, Fla.: Academic Press, 1984.

Life in Space. Boston: Little, Brown, 1983.

McAleer, Neil, *The Cosmic Mind-Boggling Book.* New

York: Warner Books, 1982.

Miller, Ron, and William K. Hartmann, *The Grand Tour: A Traveler's Guide to the Solar System.* New York: Workman, 1981.

Montoya, Earl J., and Richard O. Fimmel, *Space Pioneers and Where They Are Now.* Washington, D.C.: NASA, 1987.

Moore, Patrick:
Guide to Mars. New York: W. W. Norton, 1977.
A Guide to the Planets. New York: W. W. Norton, 1960.
The Guinness Book of Astronomy Facts & Feats. Enfield, Middlesex, England: Guinness Superlatives, 1983.
Suns, Myths and Men. London: Frederick Muller, 1968.

Moritz, Charles, ed., *Current Biography.* New York: H. W. Wilson, 1974.

Murray, Bruce, ed., *The Planets: Readings from Scientific American.* New York: W. H. Freeman, 1983.

Murray, Bruce, Michael C. Malin, and Ronald Greeley, *Earthlike Planets: Surfaces of Mercury, Venus, Earth, Moon, Mars.* San Francisco: W. H. Freeman, 1981.

Mutch, Thomas A., et al., *The Geology of Mars.* Princeton, N.J.: Princeton University Press, 1976.

National Aeronautics and Space Administration (NASA):
NASA, The First 25 Years: 1958-1983. Washington, D.C.: NASA, 1983.
Viking: The Exploration of Mars. Pasadena, Calif.: NASA, 1984.

Plant, Malcolm, *Dictionary of Space.* Essex, England: Longman House, 1986.

Ryan, Peter and Ludek Persek, *Solar Systems.* London: Penguin Books, 1978.

Sagan, Carl *Cosmos.* New York: Ballantine Books, 1980.

Schubert, Gerald, and Curt Covey, "The Atmosphere of Venus." In *The Planets,* ed. by Bruce Murray. New York: W. H. Freeman, 1983.

Short, Nicholas M., *Planetary Geology.* Englewood Cliffs, N.J.: Prentice-Hall, 1975.

Snow, Theodore P., *The Essentials of a Dynamic Universe: An Introduction to Astronomy.* St. Paul, Minn.: West, 1984.

Spitzer, Cary R., ed., *Viking Orbiter Views of Mars.* Washington, D.C.: NASA, 1980.

Strom, Robert G., *Mercury: The Elusive Planet.* Washington, D.C.: Smithsonian Institution Press, 1987.

Van Allen, James A., "Magnetospheres and the Interplanetary Medium." In *The New Solar System,* ed. by J. Kelly Beatty, Brian O'Leary, and Andrew Chaikin. Cambridge, Mass.: Sky, 1981.

Viking Lander Imaging Team, *The Martian Landscape.* Washington, D.C.: NASA, 1978.

Washburn, Mark, *Mars at Last!* New York: G. P. Putnam's Sons, 1977.

Wilson, Colin, *Starseekers.* Garden City, N.Y.: Doubleday, 1980.

Periodicals

"Bagging Elusive Mercury." *Sky & Telescope,* September 1984.

Beatty, J. Kelly, "Radar Views of Venus." *Sky & Telescope,* February 1984.

Burnham, Robert, "Venus: Planet of Fire." *Astronomy,* September 1991.

Carroll, Michael, "The Changing Face of Mars." *Astronomy,* March 1987.

Chaikin, Andrew, "Magellan Pierces Venus' Veil." *Discover,* January 1992.

Ciaccio, Edward J., "Atmospheres." *Astronomy,* May 1984.

Cintala, Mark J., James W. Head, and Thomas A. Mutch, "Characteristics of Fresh Martian Craters as a Function of Diameter: Comparison with the Moon and Mercury." *Geophysical Research Letters,* March 1976.

Cordell, Bruce M., "Mercury: The World Closest to the Sun." *Mercury,* September-October 1984.

Crossman, Ken, "Percival Lowell: A Controversial Pioneer." *Astronomy,* July 1986.

"Dr. Gerard Kuiper, Astronomer, Dies." *The New York Times,* December 25, 1973.

Driscoll, Everly, "On the Heels of Mercury." *Science News,* October 6, 1973.

Dunne, James A., "Mariner 10 Mercury Encounter." *Science,* July 12, 1974.

"Earth's Neighbors Yield Surprises." *U.S. News and World Report,* April 15, 1974.

"Exploring the Planets." *Time,* April 1, 1974.

Fink, Donald E.:
"Mariner Surpasses Mercury Flyby Goals." *Aviation Week & Space Technology,* April 8, 1974.
"Third Flyby of Mercury Yields New Data." *Aviation Week & Space Technology,* March 24, 1975.

"Focusing on Mariner 10—the Fruits of a Decade of Space Science." *Space World,* June 1975.

Freiherr, Greg, "Balloons over Venus." *Air & Space,* June-July 1988.

Gault, Donald E., et al., "Some Comparisons of Impact Craters on Mercury and the Moon." *Journal of Geophysical Research,* June 10, 1975.

Gingerich, Owen, "How Astronomers Finally Captured Mercury." *Sky & Telescope,* September 1983.

Gore, Rick:
"Between Fire and Ice: The Planets." *National Geographic,* January 1985.
"Sands of Mars." *National Geographic,* January 1977.

Guest, John E., and Donald E. Gault, "Crater Populations in the Early History of Mercury." *Geophysical Research Letters,* March 1976.

"Gyro Anomaly Alters Mariner Flyby Tactic." *Aviation Week & Space Technology,* February 4, 1974.

Haberle, Robert M., "The Climate of Mars." *Scientific American,* May 1986.

Hartmann, William K.:
"Cratering in the Solar System." *Scientific American,* January 1977.
"The Significance of the Planet Mercury." *Sky & Telescope,* May 1976.
"Infrared Thermal Studies of Mercury's Dark Side." *Sky & Telescope,* February 1971.

Kerr, Richard A.:
"Magellan: No Venusian Plate Tectonics Seen," *Science,* April 12, 1991.
"Volcanism on Mercury and the Moon, Again." *Science,* September 19, 1986.
"Where to Put the Missing Venusian Ocean." *Science,* February 21, 1986.

Kobrick, Michael, "Topography of the Terrestrial Planets." *Astronomy,* May 1982.

McCauley, John F., et al., "Stratigraphy of the Caloris Basin, Mercury." *Icarus 47,* 1981, pages 184-202.

Malin, Michael C., "Observations of Intercrater Plains on Mercury." *Geophysical Research Letters,* October 1976.

"Mariner Experiments to Give First Details of Planetary Phenomena." *Aviation Week & Space Technology,* October 8, 1973.

"Mariner on Course Following First Correction Maneuver." *Aviation Week & Space Technology,* November 26, 1973.

"Mariner-10." *Space World,* January 1974.

"Mariner 10 Returns to Mercury." *Science News,* September 14, 1974.

"Mariner 10's Cameras Cold but Functional." *Science News,* November 17, 1973.

"Mariner 10 Set for Second Look at Mercury." *Space World,* February 1975.

"Mars as Viking Sees It." *National Geographic,* January 1977.

"Mercury: Pocked Planet Gives Up Its Secrets." *U.S. News and World Report,* December 16, 1974.

"Mercury: Scarps, Massive Impact Craters and a Compressed Core." *Science News,* September 28, 1974.

"Mercury: Surface Composition from the Reflection Spectrum." *Science,* November 17, 1972.

"Mercury at Last." *Science News,* March 30, 1974.

"Mercury Revisited by Mariner 10." *Sky & Telescope,* May 1975.

"Mercury's Atmosphere Is Rich in Sodium." *Sky & Telescope,* November 1985.

"Mercury's Double Dawn." *Time,* March 10, 1967.

"Mercury's Magnetism." *Time,* March 31, 1975.

"Mercury's Magnetism Is Its Own." *Science News,* March 22, 1975.

"Mission to Mercury." *Newsweek,* April 8, 1974.

Montoya, Earl J., and Richard O. Fimmel, "The First Pioneers." *Mercury,* March-April 1988.

"More from Mercury." *Scientific American,* September 1974.

Nozette, Stewart, and Peter Ford, "Venus: A World Revealed." *Astronomy,* March 1981.

"Obituaries: Gerard P. Kuiper." *Physics Today,* March 1974.

"Obituaries: Giuseppe Colombo." *Physics Today,* October 1984.

Overbye, Dennis, "Admiral of the Solar System." *Discover,* November 1981.

Pettengill, G. H., et al., "Magellan: Radar Performance and Data Products." *Science,* April 12, 1991.

Pollack, James B., "Origin and History of the Outer Planets." *Annual Review of Astronomy & Astrophysics,* 1984, pages 389-424.

Potter, Andrew E., and Thomas H. Morgan:
"Discovery of Sodium in the Atmosphere of Mercury." *Science,* August 16, 1985.
"Potassium in the Atmosphere of Mercury." *Icarus 67,* 1986, pages 336-340.
"Variation of Sodium on Mercury with Solar Radiation Pressure." *Icarus 71,* 1987, pages 472-477.

Prinn, Ronald G., "The Volcanoes and Clouds of Venus." *Scientific American,* March 1985.

"Radar Map of Mercury: Craters, Rolling Hills." *Science News,* February 2, 1974.

Sagdeev, Roald Z., and Albert A. Galeev, "Comet Halley and the Solar Wind." *Sky & Telescope,* March 1987.

Saunders, R. S., and G. H. Pettengill, "Magellan: Mission Summary." *Science,* April 12, 1991.

Saunders, R. S., et al., "An Overview of Venus Geology." *Science,* April 12, 1991.

Schefter, Jim:
"Pioneer Venus—Report from the Mysterious Planet." *Popular Science,* April 1979.
"Survival on Venus: How the Russians Did It." *Popular Science,* November 1982.

Schultz, P. H., "Experimental Planetary Impact Research." *International Journal of Impact Engineering,* 1987, pages 569-576.

Shemansky, Don E., "Revised Atmospheric Species Abundances at Mercury: The Debacle of Bad g Values." *The Mercury Messenger,* August 1988.

"Sidereal Messenger." *Scientific American,* May 1974.

Solomon, Sean, "Some Aspects of Core Formation in Mercury." *Icarus 28,* 1976, pages 509-521.

Solomon, Sean, and J. W. Head, "Fundamental Issues in the Geology and Geophysics of Venus." *Science,* April 12, 1991.

"Space Detective Stops a Suicide." *Science News,* May 18, 1974.

"The Strange and Cratered World of Mercury." *Science News,* April 6, 1974.

Sullivan, Walter:
"Mariner 10 Will Visit Mercury Twice." *The New York Times,* March 28, 1974.
"Mercury's Mysterious 'Moon' Turns Out to Be a Star." *The New York Times,* April 2, 1974.
"A Possible Moon of Mercury Is Detected." *The New York Times,* April 1, 1974.

Thomas, Pierre G., and Philippe Masson, "Tectonics of the Caloris Area on Mercury: An Alternative View." *Icarus 58,* 1984, pages 396-402.

Trask, Newell J., and John E. Guest, "Preliminary Geologic Terrain Map of Mercury." *Journal of Geophysical Research,* June 10, 1975.

Trask, Newell J., and Robert G. Strom, "Additional Evidence of Mercurian Volcanism." *Icarus 28,* 1976, page 559.

"Venus Past: Wet or Dry?" *Sky & Telescope,* April 1988.

Von Braun, Wernher, "The Never-before-Seen Face of Mercury." *Popular Science,* August 1974.

Weaver, Kenneth F.:
"Journey to Mars." *National Geographic,* February 1973.
"Mariner Unveils Venus and Mercury." *National Geographic,* June 1975.

Weidenschilling, S. J., "Iron/Silicate Fractionation and the Origin of Mercury." *Icarus 35,* 1978, pages 99-111.

Wilhelms, Don E., "Mercurian Volcanism Questioned." *Icarus 28,* 1976, pages 551-558.

"With Mariner 10 en Route to Mercury." *Science News,* December 15, 1973.

Other Publications

"Gravity Assist." NASA Educational Brief. Washington, D.C.: U.S. Government Printing Office, 1985.

Young, Carolyn, ed., *The Magellan Venus Explorer's Guide,* NASA/JPL, August 1, 1990.

INDEX

ACKNOWLEDGMENTS

The editors wish to thank Ferdinand Anton, Munich; Barbara Chappell, U.S. Geological Survey, Reston, Va.; Steven Dick, U.S. Naval Observatory, Washington, D.C.; Marie-Pierre Dion, Conservateur de la Bibliothèque Municipale de Valenciennes, France; Rainer Herbster, Deutsches Museum, Munich; Peter Hingley, Royal Astronomical Society, London; Ralph Kahn Jet Propulsion Laboratory, Pasadena, Calif.; Vanessa Myers, NASA Goddard Space Flight Center, Greenbelt, Md.; Stanton Peale, University of California, Santa Barbara, Calif.; Jeffrey Plescia, Jet Propulsion Laboratory, Pasadena, Calif.; Jim Reardon, NASA Goddard Space Flight Center, Greenbelt, Md.; Craig Roberts, Computer Science Corporation, Seabrook, Md.; Patricia Ross, NASA Goddard Space Flight Center, Greenbelt, Md.; C. T. Russell, University of California, Los Angeles; Peter Schultz, Brown University, Providence, R.I.; Trudy Sinnott, U.S. Geological Survey, Reston, Va.; Kathy Teague, U.S. Geological Survey, Flagstaff, Ariz.; Bob Tice, NASA Goddard Space Flight Center, Greenbelt, Md.; Richard Zurek, Jet Propulsion Laboratory, Pasadena, Calif.

PICTURE CREDITS

Time-Life Books is a division of Time Life Inc., a wholly owned subsidiary of
THE TIME INC. BOOK COMPANY

TIME-LIFE BOOKS

PRESIDENT: Mary N. Davis

Managing Editor: Thomas H. Flaherty
Director of Editorial Resources:
Elise D. Ritter-Clough
Director of Photography and Research:
John Conrad Weiser
Editorial Board: Dale M. Brown, Roberta Conlan, Laura Foreman, Lee Hassig, Jim Hicks, Blaine Marshall, Rita Thievon Mullin, Henry Woodhead
Assistant Director of Editorial Resources/ Training Manager: Norma E. Shaw

PUBLISHER: Robert H. Smith

Associate Publisher: Trevor Lunn
Editorial Director: Donia Steele
Marketing Director: Regina Hall
Production Manager: Marlene Zack
Supervisor of Quality Control: James King

Editorial Operations
Production: Celia Beattie
Library: Louise D. Forstall
Computer Composition: Deborah G. Tait (Manager), Monika D. Thayer, Janet Barnes Syring, Lillian Daniels
Interactive Media Specialist: Patti H. Cass

Correspondents: Elisabeth Kraemer-Singh (Bonn), Maria Vincenza Aloisi (Paris), Ann Natanson (Rome). Valuable assistance was also provided by Christine Hinze (London), Ann Wise (Rome).

VOYAGE THROUGH THE UNIVERSE

SERIES DIRECTOR: Roberta Conlan
Series Administrator: Judith W. Shanks

Editorial Staff for *The Near Planets*
Designer: Dale Pollekoff
Associate Editor: Sally Collins (pictures)
Text Editors: Pat Daniels (principal), Peter Pocock
Researchers: Tina S. McDowell, Barbara C. Mallen, Barbara Sause
Writer: Esther Ferington
Assistant Designer: Brook Mowrey
Editorial Assistant: Jayne A. L. Dover
Copy Coordinator: Darcie Conner Johnston
Picture Coordinators: Richard Karno, Bob Wooldridge

Special Contributors: Joseph Anthony, J. Kelly Beatty, Tom Burroughs, Roberta Friedman, Peter Gwynne, Leon Jaroff, Gina Maranto, Gregory A. Mock, Eugene Rodgers, Chuck Smith, M. Mitchell Waldrop, Mark Washburn (text); Lynn Cook, Pia Farrell, Marilyn Fenischel, Jeff Kenney, Hugh McIntosh, Eugenia S. Scharf, Jacqueline Shaffer, Joann Stern, Carolyn Tozier, Naida Maris Yolen (research); Michael Kalen Smith (index).

CONSULTANTS

RAYMOND M. BATSON is chief of the Planetary Cartography Section of the U.S. Geological Survey in Flagstaff, Arizona. He has participated as a coinvestigator in several planetary missions.

MICHAEL H. CARR, a planetary geologist with the U.S. Geological Survey in Menlo Park, California, specializes in the terrestrial planets, including the study of Martian terrain. He is a member of the Voyager and Galileo imaging teams.

JAMES DUNNE is project manager at the Jet Propulsion Laboratory for the Soviet Phobos mission to Mars. Much of his work involves the use of x-rays to conduct remote analysis of planetary surfaces.

TED A. MAXWELL is chairman of the Center for Earth and Planetary Studies, National Air and Space Museum, Smithsonian Institution. He is also director of the Smithsonian's Regional Planetary Image Facility.

RONALD G. PRINN teaches at Massachusetts Institute of Technology. His research specialty is planetary atmospheres, including the weather on Venus.

P. KENNETH SEIDELMANN is director of the Nautical Almanac Office of the U.S. Naval Observatory, Washington, D.C.

ROBERT STROM is with the Lunar and Planetary Laboratory at the University of Arizona. A member of the imaging teams for *Mariner 10* and Voyager, he specializes in the geology and geophysics of Mercury.

JOSEPH N. TATAREWICZ, historian of astronomy, is Associate Curator in the Space Science and Exploration Department of the National Air and Space Museum of the Smithsonian Institution.

MARIA T. ZUBER is a geophysicist at the NASA Goddard Space Flight Center. Her research on the tectonic structures of the terrestrial planets is based on theoretical modeling and the interpretation of spacecraft data.

Library of Congress Cataloging in Publication Data
The Near planets/by the editors of Time-Life Books.
    p. cm. —(Voyage through the universe)
    Includes bibliographical references (p.   )
and index.
    ISBN 0-8094-9066-8
    ISBN 0-8094-9067-6 (lib. bdg.)
    1. Inner planets. 2. Mars (Planet).
3. Mercury (Planet). 4. Venus (Planet).
I. Time-Life Books. II. Series.
QB601.N4 1992
523.4—dc20   92-8443
                                             CIP

For information on and a full description of any of the Time-Life Books series, please call 1-800-621-7026 or write:
Reader Information
Time-Life Customer Service
P.O. Box C-32068
Richmond, Virginia 23261-2068

REVISIONS STAFF

EDITOR: Roberta Conlan

Associate Editor/Research: Quentin G. Story
Art Director: Robert K. Herndon
Revisions Coordinator: Barbara Fairchild Quarmby
Picture Coordinator: David Beard
Assistant Art Director: Kathleen Mallow

Consultant: Sean C. Solomon, a professor of geophysics at the Massachusetts Institute of Technology, specializes in earthquake seismology, marine geophysics, and planetary geology. He is a member of the Magellan science team.